Lecture Notes in Mathematics 1933

Editors:
J.-M. Morel, Cachan
F. Takens, Groningen
B. Teissier, Paris

T0222662

Miodrag Petković

Point Estimation of Root Finding Methods

 Springer

Miodrag Petković
Department of Mathematics
Faculty of Electronic Engineering
University of Niš
A. Medvedeva 14
P.O. Box 73
18000 Niš
Serbia
msp@junis.ni.yu
msp@eunet.yu
www.miodragpetkovic.com

ISBN: 978-3-540-77850-9   e-ISBN: 978-3-540-77851-6
DOI: 10.1007/978-3-540-77851-6

Lecture Notes in Mathematics ISSN print edition: 0075-8434
                    ISSN electronic edition: 1617-9692

Library of Congress Control Number: 2008921391

Mathematics Subject Classification (2000): 65H05

Cover design: WMXDesign GmbH

Printed on acid-free paper

9 8 7 6 5 4 3 2 1

springer.com

To Ljiljana

Preface

The problem of solving nonlinear equations and systems of equations ranks among the most significant in the theory and practice, not only of applied mathematics but also of many branches of engineering sciences, physics, computer science, astronomy, finance, and so on. A glance at the bibliography and the list of great mathematicians who have worked on this topic points to a high level of contemporary interest. Although the rapid development of digital computers led to the effective implementation of many numerical methods, in practical realization, it is necessary to solve various problems such as computational efficiency based on the total central processor unit time, the construction of iterative methods which possess a fast convergence in the presence of multiplicity (or clusters) of a desired solution, the control of rounding errors, information about error bounds of obtained approximate solution, stating computationally verifiable initial conditions that ensure a safe convergence, etc. It is the solution of these challenging problems that was the principal motivation for the present study.

In this book, we are mainly concerned with the statement and study of initial conditions that provide the guaranteed convergence of an iterative method for solving equations of the form $f(z) = 0$. The traditional approach to this problem is mainly based on asymptotic convergence analysis using some strong hypotheses on differentiability and derivative bounds in a rather wide domain. This kind of conditions often involves some unknown parameters as constants, or even desired roots of equation in the estimation procedure. Such results are most frequently of theoretical importance and they provide only a qualitative description of the convergence property. The first results dealing with the computationally verifiable domain of convergence were obtained by Smale (1981), Smale (1986), Shub and Smale (1985), and Kim (1985). This approach, often referred to as "point estimation theory," treats convergence conditions and the domain of convergence in solving an equation $f(z) = 0$ using only the information of f at the initial point $z^{(0)}$.

In 1981, Smale introduced the concept of an *approximate zero* as an initial point which provides the safe convergence of Newton's method. Later, in 1986, he considered the convergence of Newton's method from data at a single point. X. Wang and Han (1989) and D. Wang and Zhao (1995) obtained some improved results. The study in this field was extended by Kim (1988) and Curry (1989) to some higher-order iterative methods including Euler's method and Halley's method, and by Chen (1989), who dealt with the general Newton-like quadratically convergent iterative algorithms. A short review of these results is given in the first part of Chap. 2. Wang–Zhao's improvement of Smale's convergence theorem and an interesting application to the Durand–Kerner method for the simultaneous determination of polynomial zeros are presented in the second part of Chap. 2.

The main aim of this book is to state such quantitative initial conditions for predicting the immediate appearance of the guaranteed and fast convergence of the considered numerical algorithm. Special attention is paid to the convergence analysis of iterative methods for the simultaneous determination of the zeros of algebraic polynomials. However, the problem of the choice of initial approximations which ensure a safe convergence is a very difficult one and it cannot be solved in a satisfactory way in general, not even in the case of simple functions, such as algebraic polynomials. In 1995, the author of this book and his contributors developed two procedures to state initial conditions for the safe convergence of simultaneous methods for finding polynomial zeros. The results were based on suitable localization theorems for polynomial zeros and the convergence of error sequences. Chapter 3 is devoted to initial conditions for the guaranteed convergence of most frequently used iterative methods for the simultaneous approximations of all simple zeros of algebraic polynomials. These conditions depend only on the coefficients of a given polynomial $P(z) = z^n + a_{n-1}z^{n-1} + \cdots + a_1 z + a_0$ of degree n and the vector of initial approximations $z^{(0)} = \left(z_1^{(0)}, \ldots, z_n^{(0)}\right)$. In particular, some efficient a posteriori error bound methods that produce disks containing the sought zeros and require fewer numerical operations than the corresponding ordinary interval methods are considered in the last part of Chap. 3.

The new results presented in Chaps. 4 and 5 are concerned with the higher-order families of methods for the simultaneous determination of complex zeros. These methods are based on the iterative formula of Hansen–Patrick's type for finding a single zero. As in Chap. 3, we state computationally verifiable initial conditions that guarantee the convergence of the presented methods. Initial conditions ensuring convergence of the corresponding iterative methods for the inclusion of polynomial zeros are established in Chap. 5. Convergence behavior of the considered methods is illustrated by numerical examples.

I wish to thank Professor C. Carstensen of Humboldt University in Berlin. Our joint work (*Numer. Math.* 1995) had a stimulating impact on the development of the basic ideas for obtaining some results given in this book. I am grateful to Professor S. Smale, the founder of the point estimation theory, who drew my attention to his pioneering work. I am also thankful to my contributors and coauthors of joint papers Professor T. Sakurai of the

University of Tsukuba, Professor Đ. Herceg of the University of Novi Sad, Professor S. Ilić, Dr. L. Rančić, Dr. D. Milošević, and Professor D. Đorđević of the University of Niš for numerous important suggestions and valuable comments. What I especially wish to acknowledge is the assistance and exceptional efforts of Sonja Dix-Stojanović and Aleksandra Milošević who read the complete manuscript. Many small errors were eliminated in this manner.

My principal thanks, however, go to my wife Professor Ljiljana Petković for her never-failing support, encouragement, and permanent discussions during the preparation of the manuscript.

University of Niš,   *Miodrag S. Petković*
Faculty of Electronic Engineering,
Department of Mathematics,
Niš 18000, Serbia
December 2007

Contents

Chapter 1
Basic Concepts

In this chapter, we give some basic concepts and properties, necessary in our investigation of convergence characteristics of root finding methods. Most of these methods are reviewed in Sect. 1.1, together with some historical notes and various principles for their construction. Section 1.2 contains several results concerning the localization of polynomial zeros. We restrict ourselves to inclusion disks in the complex plane that contain complex zeros of a given polynomial. In Sect. 1.3, we give the basic properties and operations of circular complex interval arithmetic, prerequisite to a careful analysis of the bounds of complex quantities that appear in our study and the construction of inclusion methods described in Sect. 5.3.

1.1 Simultaneous Methods for Finding Polynomial Zeros

The problem of determining the zeros of a given polynomial is one of the first nonlinear problems that mathematicians meet in their research and practice. Although this problem seems to be simple at first sight, a perfect algorithm for finding polynomial zeros has not been established yet, in spite of numerous algorithms developed during the last 40 years. Each numerical method possesses its own advantages and disadvantages, so that it is not easy to choose the "best" method for a given polynomial equation. Let us emphasize that the rapid development of computing machines implies that many algorithms, formerly of academic interest only, become feasible in practice.

Most algorithms calculate only one zero at a time. In cases when all zeros are needed, these algorithms usually work serially as follows: when a zero has been computed to sufficient accuracy, then the corresponding linear factor is removed from the polynomial by the Horner scheme and the process is applied again to determine a zero of the "deflated" polynomial whose degree is now lowered by one. This is the method of successive *deflations*. If a great accuracy of desired approximations to the zeros is required, the polynomial obtained

M. Petković, *Point Estimation of Root Finding Methods.* Lecture Notes in Mathematics 1933,
© Springer-Verlag Berlin Heidelberg 2008

after divisions by the previously calculated (inaccurate) linear factors may be falsified to an extent which makes the remaining approximate zeros erroneous. This is a flaw of the method of successive removal of linear factors. The next disadvantage appears in those situations where it is sufficient to find approximations with only a few significant digits. But, as mentioned above, the method of deflation requires approximations of great accuracy. Besides, this procedure cannot ensure that the zeros are determined in increasing order of magnitude (see Wilkinson [190]), which is an additional shortcoming of deflation.

The above difficulties can be overcome in many situations by approximating all zeros simultaneously. Various approaches to these procedures have been developed: the method of search and exclusion (Henrici [57, Sect. 6.11]), methods based on the fixed point relations (e.g., Börsch-Supan [9], [10], Ehrlich [33], X. Wang and Zheng [182], Gargantini [47], [48]), qd algorithm (Henrici [57, Sect. 7.6]), a globally convergent algorithm that is implemented interactively (Farmer and Loizou [39]), tridiagonal matrix method (Brugnano and Trigiante [12], Schmeisser [159]), companion matrix methods (Smith [168], Niu and Sakurai [93], Fiedler [40], Malek and Vaillancourt [86]), methods based on the application of root finders to a suitable function ([69], [124], [146], [156]), methods which use rational approximations (Carstensen and Sakurai [18], Sakurai et al. [157], [158]), and others (see, for instance, Wilf [189], Pasquini and Trigiante [103], Jankins and Traub [67], Farmer and Loizou [37], [38]). See also Pan's survey paper [101] and references cited therein.

Part I: Simultaneous Methods Based on Fixed Point Relations

In this book, we deal mainly with the simultaneous methods based on fixed point relations (FPR). Such an approach generates algorithms with very fast convergence in complex "point" arithmetic as well as in complex interval arithmetic using the following procedure.

Let $\zeta_1, \ldots, \zeta_n$ be the zeros of a given monic (normalized, highest coefficient 1) polynomial $P(z) = z^n + a_{n-1} z^{n-1} + \cdots + a_1 z + a_0$ of degree n and let $z_1, \ldots, z_n$ be their respective approximations. We consider two types of FPR

$$\zeta_i = F_1(z_1, \ldots, z_{i-1}, z_i, \zeta_i, z_{i+1}, \ldots, z_n), \tag{1.1}$$

$$\zeta_i = F_2(\zeta_1, \ldots, \zeta_{i-1}, z_i, \zeta_{i+1}, \ldots, \zeta_n), \tag{1.2}$$

where $i \in \boldsymbol{I}_n := \{1, \ldots, n\}$. Now we give several FPR which have been the basis for the construction of the most frequently used iterative methods for the simultaneous determination of polynomial zeros in complex arithmetic and complex interval arithmetic. In the latter development, we will frequently use *Weierstrass' correction* $W_i(z_i) = P(z_i) / \prod_{j \neq i}(z_i - z_j)$ $(i \in \boldsymbol{I}_n)$. Sometimes, we will write W_i instead of $W_i(z_i)$. In addition to the references given behind the type of FPR, the derivation of these FPR may be found in the book [109] of M. Petković.

For brevity, we will sometimes write

$$\sum_{j \neq i} x_j \text{ instead of } \sum_{\substack{j=1 \\ j \neq i}}^{n} x_j \text{ and } \prod_{j \neq i} x_j \text{ instead of } \prod_{\substack{j=1 \\ j \neq i}}^{n} x_j.$$

Example 1.1. **The Weierstrass-like FPR** [109]:

$$\zeta_i = z - \frac{P(z)}{\displaystyle\prod_{\substack{j=1 \\ j \neq i}}^{n}(z - \zeta_j)} \quad (i \in \boldsymbol{I}_n). \tag{F_1}$$

Equation (F_1) follows from the factorization $P(z) = \displaystyle\prod_{j=1}^{n}(z - \zeta_j) = (z - \zeta_i)\displaystyle\prod_{\substack{j=1 \\ j \neq i}}^{n}(z - \zeta_j).$

Example 1.2. **The Newton-like FPR** [50], [106]:

$$\zeta_i = z - \frac{1}{\dfrac{P'(z)}{P(z)} - \displaystyle\sum_{\substack{j=1 \\ j \neq i}}^{n}\dfrac{1}{z - \zeta_j}} \quad (i \in \boldsymbol{I}_n). \tag{F_2}$$

Applying the logarithmic derivative to $P(z) = \displaystyle\prod_{j=1}^{n}(z - \zeta_j)$, the identity

$$\frac{P'(z)}{P(z)} = \sum_{j=1}^{n}\frac{1}{z - \zeta_j} \tag{1.3}$$

is obtained. Finding $z - \zeta_i$ from (1.3), we get (F_2).

Example 1.3. **The Börsch-Supan-like FPR** [11], [107]:

$$\zeta_i = z - \frac{W_i}{1 + \displaystyle\sum_{\substack{j=1 \\ j \neq i}}^{n}\dfrac{W_j}{\zeta_i - z_j}} \quad (i \in \boldsymbol{I}_n). \tag{F_3}$$

Lagrange's interpolation applied to the distinct points $z_1, \ldots, z_n$ ($\neq \zeta_i$, $i \in \boldsymbol{I}_n$) gives

$$P(z) = W_i \prod_{\substack{j=1 \\ j \neq i}}^{n} (z - z_j) + \prod_{j=1}^{n} (z - z_j) \left(\sum_{\substack{j=1 \\ j \neq i}}^{n} \frac{W_j}{z - z_j} + 1 \right). \qquad (1.4)$$

Taking $z = \zeta_i$ and solving the obtained equation in $\zeta_i - z_i$, from (1.4) we derive (F_3).

Example 1.4. **The square root FPR** [47], [106]:

$$\zeta_i = z - \frac{1}{\left[\dfrac{P'(z)^2 - P(z)P''(z)}{P(z)^2} - \displaystyle\sum_{\substack{j=1 \\ j \neq i}}^{n} \dfrac{1}{(z - \zeta_j)^2} \right]^{1/2}} \qquad (i \in \boldsymbol{I}_n). \qquad (F_4)$$

Differentiation of the identity (1.3) yields

$$-\left(\frac{P'(z)}{P(z)} \right)' = \frac{P'(z)^2 - P(z)P''(z)}{P(z)^2} = \sum_{j=1}^{n} \frac{1}{(z - \zeta_j)^2}, \qquad (1.5)$$

wherefrom we extract the term $(z - \zeta_i)^2$ and derive (F_4).

Example 1.5. **The Halley-like FPR** [109], [182]:

$$\zeta_i = z - \frac{1}{\dfrac{P'(z)}{P(z)} - \dfrac{P''(z)}{2P'(z)} - \dfrac{P(z)}{2P'(z)} \left[\left(\displaystyle\sum_{\substack{j=1 \\ j \neq i}}^{n} \dfrac{1}{z - \zeta_j} \right)^2 + \displaystyle\sum_{\substack{j=1 \\ j \neq i}}^{n} \dfrac{1}{(z - \zeta_j)^2} \right]}. \qquad (F_5)$$

Equation (F_5) can be obtained by substituting the sums (1.3) and (1.5) in the relation

$$\frac{P''(z)}{P(z)} = \left(\frac{P'(z)}{P(z)} \right)^2 + \left(\frac{P'(z)}{P(z)} \right)'.$$

Actually, (F_5) is a special case of a general fixed point relation derived by X. Wang and Zheng [182] by the use of Bell's polynomials (see Comment (M_5)).

Substituting the exact zeros $\zeta_1, \ldots, \zeta_n$ by their respective approximations $z_1, \ldots, z_n$ and putting $z = z_i$, from (1.1) and (1.2), we obtain iterative schemes

$$\hat{z}_i = F_1(z_1, \ldots, z_n) \quad (i \in \boldsymbol{I}_n), \qquad (1.6)$$
$$\hat{z}_i = F_2(z_1, \ldots, z_n) \quad (i \in \boldsymbol{I}_n), \qquad (1.7)$$

in (ordinary) complex arithmetic, where $\hat{z}_i$ is a new approximation to the zero ζ_i. Another approach consisting of the substitution of the zeros on the right side of (1.1) and (1.2) by their inclusion disks enables the construction of interval methods in circular complex interval arithmetic (see Sects. 1.3 and 5.3).

For illustration, we list below the corresponding simultaneous iterative methods based on the FPR given in Examples 1.1–1.5 and having the form (1.6) or (1.7).

The Durand–Kerner's or Weierstrass' method [1], [30], [32], [72], [148], [187], order 2:

$$\hat{z}_i = z_i - W_i = z_i - \frac{P(z_i)}{\displaystyle\prod_{\substack{j=1 \\ j \neq i}}^{n} (z_i - z_j)} \quad (i \in \boldsymbol{I}_n). \tag{M_1}$$

The Ehrlich–Aberth's method [1], [31], [33], [85], order 3:

$$\hat{z}_i = z_i - \frac{1}{\dfrac{P'(z_i)}{P(z_i)} - \displaystyle\sum_{\substack{j=1 \\ j \neq i}}^{n} \dfrac{1}{z_i - z_j}} \quad (i \in \boldsymbol{I}_n). \tag{M_2}$$

The Börsch-Supan's method [10], [95], order 3:

$$\hat{z}_i = z_i - \frac{W_i}{1 + \displaystyle\sum_{\substack{j=1 \\ j \neq i}}^{n} \dfrac{W_j}{z_i - z_j}} \quad (i \in \boldsymbol{I}_n). \tag{M_3}$$

Let us introduce $\delta_{k,i} = \dfrac{P^{(k)}(z_i)}{P(z_i)}$ $(k = 1, 2)$. Then

$$\delta_{1,i}^2 - \delta_{2,i} = \frac{P'(z_i)^2 - P(z_i)P''(z_i)}{P(z_i)^2}.$$

The square root method [47], [142], order 4:

$$\hat{z}_i = z_i - \frac{1}{\left[\delta_{1,i}^2 - \delta_{2,i} - \displaystyle\sum_{\substack{j=1 \\ j \neq i}}^{n} \dfrac{1}{(z_i - z_j)^2} \right]^{1/2}} \quad (i \in \boldsymbol{I}_n). \tag{M_4}$$

The Halley-like or Wang–Zheng's method [182], order 4:

$$\hat{z}_i = z_i - \frac{1}{f(z_i) - \dfrac{P(z_i)}{2P'(z_i)} \left[\left(\displaystyle\sum_{\substack{j=1 \\ j \neq i}}^{n} \dfrac{1}{z_i - z_j} \right)^2 + \displaystyle\sum_{\substack{j=1 \\ j \neq i}}^{n} \dfrac{1}{(z_i - z_j)^2} \right]} \quad (i \in \boldsymbol{I}_n),$$

$$\tag{M_5}$$

where

$$f(z_i) = \frac{P'(z_i)}{P(z_i)} - \frac{P''(z_i)}{2P'(z_i)} \tag{1.8}$$

is the denominator of Halley's correction

$$H(z_i) = H_i = \frac{1}{f(z_i)}, \tag{1.9}$$

which appears in the well-known classical Halley's method [4], [45], [54]

$$\hat{z}_i = z_i - H(z_i) = z_i - \frac{1}{\dfrac{P'(z_i)}{P(z_i)} - \dfrac{P''(z_i)}{2P'(z_i)}}. \tag{1.10}$$

Comment (M_1). Formula (M_1) has been rediscovered several times (see Durand [32], Dochev [30], Börsch-Supan [9], Kerner [72], M. Prešić [147], S. B. Prešić [149]) and it has been derived in various ways. But we emphasize the little known fact that this formula was known seven decades ago. In his lecture on the session of König, Academy of Science, held on 17 December 1891, Weierstrass communicated a new constructive proof of the fundamental theorem of algebra (printed in [187]). In this proof, Weierstrass used the sequences of numerical entries $\{a_\nu^{(\lambda)}\}$ ($\nu = 1, \ldots, n$, $\lambda = 0, 1, 2, \ldots$) defined successively by (eq. (29) in Weierstrass' work [187])

$$\begin{cases} a'_\nu = a_\nu - \dfrac{P(a_\nu)}{\prod_\mu (a_\nu - a_\mu)} \\ a''_\nu = a'_\nu - \dfrac{P(a'_\nu)}{\prod_\mu (a'_\nu - a'_\mu)} \quad (\nu = 1, \ldots, n,\ \mu \gtrless \nu), \\ a'''_\nu = a''_\nu - \dfrac{P(a''_\nu)}{\prod_\mu (a''_\nu - a''_\mu)} \\ \text{and so on} \end{cases} \tag{1.11}$$

where P is a polynomial of degree n with the zeros $x_1, \ldots, x_n$.

The proof of the quadratic convergence of the iterative method (M_1) is ascribed to Dochev [30], although his proof is not quite precise. The works [82], [148], [162] offer a more precise proof. But it seems that the quadratic convergence of the sequence (1.11) was known to Weierstrass. Namely, for the maximal absolute differences $\varepsilon^{(\lambda)} = \max\limits_{1 \le \nu \le n} |a_\nu^{(\lambda)} - x_\nu|$, he derived the following inequality (eq. (32) in [187])

$$\varepsilon^{(\lambda)} < \left(\varepsilon^{(0)}\right)^{2^\lambda} \quad (\lambda = 1, 2, \ldots),$$

which points to the quadratic convergence of the sequences $\{a_\nu^{(\lambda)}\}$.

Note that Weierstrass did not use (1.11) for the numerical calculation of polynomial zeros. Durand [32] and Dochev [30] were the first to apply the iterative formula (M_1) in practice for the simultaneous approximation of polynomial zeros.

In 1966, Kerner [72] proved that (M_1) is, in fact, Newton's method $\hat{z} = z - F'(z)^{-1}F(z)$ for solving nonlinear systems applied to the system of nonlinear equations (known as Viète's formulae)

$$(-1)^k \varphi_k(z_1, \ldots, z_n) - a_k = 0, \quad (k = 1, \ldots, n), \tag{1.12}$$

where φ_k denotes the kth elementary symmetric function:

$$\varphi_k = \sum_{1 \le j_1 \le \cdots j_k \le n} z_{j_1} z_{j_2} \cdots z_{j_k}.$$

Since Newton's method is quadratically convergent, it follows immediately that the iterative method (M_1) also has quadratic convergence.

The iterative method (M_1) shares with the Halley's method (1.10) the distinction of being the most frequently rediscovered method in the literature. From the fact that many authors dealt with the formula (M_1), the iterative method (M_1) is called Weierstrass', Durand–Kerner's, or Weierstrass–Dochev's method; other combinations also appear in literature.

According to a great number of numerical experiments, many authors have conjectured that the method (M_1) possesses a global convergence in practice for almost all starting vectors $\boldsymbol{z}^{(0)} = (z_1^{(0)}, \ldots, z_n^{(0)})$, assuming that the components of $\boldsymbol{z}^{(0)}$ are disjoint. This was proved for $n = 2$ (see [52], [64]) and for the cubic polynomial $P(z) = z^3$ (Yamagishi [64]), but this is an open problem still for a general $n \ge 3$.

Let us note that the method (M_1) works well even for the case where the zeros of P are not necessarily distinct (see Fraigniaud [43], Miyakoda [89], Pasquini and Trigiante [103], Carstensen [15], Kyurkchiev [81], Yamamoto, Furakane, and Nogura [193], Kanno, Kyurkchiev, and Yamamoto [69], Yamamoto, Kanno, and Atanassova [194], etc.). For these excellent properties and great computational efficiency, this method is one of the most frequently used simultaneous methods for determining polynomial zeros (see [109, Chap. 6]).

Comment (M_2). Although the method (M_2) was first suggested by Maehly [85] in 1954 for a refinement of the Newton's method and used by Börsch-Supan [9] in finding a posteriori error bounds for the zeros of polynomials, it is more often referred to as the Ehrlich–Aberth's method. Ehrlich [33] proved the cubic convergence of this method and Aberth [1] gave important contribution in its practical realization. The method (M_2) can also be derived from the Halley's method (1.10) using the approximation (see (1.49))

$$\frac{P''(z_i)}{2P'(z_i)} \approx \sum_{\substack{j=1 \\ j \neq i}}^{n} \frac{1}{z_i - z_j}$$

(e.g., [37], [145]).

The Ehrlich–Aberth's method (M_2) can be easily accelerated by replacing ζ_j with Newton's approximation $z_j - P(z_i)/P'(z_i)$ in (F_2) instead of the current approximation z_j. In this way, we obtain

The Ehrlich–Aberth's method with Newton's corrections [97], order 4:

$$\hat{z}_i = z_i - \frac{1}{\dfrac{1}{N_i} - \sum_{\substack{j=1 \\ j \neq i}}^{n} \dfrac{1}{z_i - z_j + N_j}}, \quad N_j = \frac{P(z_j)}{P'(z_j)} \quad (i \in \boldsymbol{I}_n). \qquad (M_6)$$

For computational purpose, it is preferable to calculate Newton's approximations $z_j - N_j$ in advance, before starting the iteration. Comparing (M_2) and (M_6), we observe that Newton's corrections can themselves be used to improve the convergence rate; the increase of the convergence order from 3 to 4 is attained by using a negligible number of additional operations. For this reason, the method (M_6) is one of the most efficient methods for the simultaneous determination of polynomial zeros.

Comment (M_3). As far as we know, the method (M_3) was first suggested by Börsch-Supan [10] for the estimation of approximations to the zeros of polynomials and their computation. The method (M_3) was later considered by Nourein [96] and Werner [188].

Carstensen [15] noted that the iterative formulae (M_2) and (M_3) are equivalent. This follows according to the identity

$$\frac{P'(z_i)}{P(z_i)} - \sum_{\substack{j=1 \\ j \neq i}}^{n} \frac{1}{z_i - z_j} = \frac{1}{W_i} \left(1 + \sum_{\substack{j=1 \\ j \neq i}}^{n} \frac{W_j}{z_i - z_j} \right),$$

which is proved in Sect. 3.3. Nevertheless, these formulae do not produce the same results in practice due to the rounding errors appearing when arithmetic of finite precision is used.

In a similar way as in the case of the Ehrlich–Aberth's method (M_2), we can accelerate the Börsch-Supan's method (M_3), substituting ζ_i in (F_3) directly with Weierstrass' approximation $z_i - W_i$; in this manner, we construct

The Börsch-Supan's method with Weierstrass' correction [95], [96], order 4:

$$\hat{z}_i = z_i - \frac{W_i}{1 + \sum_{\substack{j=1 \\ j \neq i}}^{n} \dfrac{W_j}{z_i - W_i - z_j}} \quad (i \in \boldsymbol{I}_n). \qquad (M_7)$$

The improved method (M_7) possesses a great computational efficiency since the increase of the convergence order requires only a few additional calculations. This method was proposed by Nourein [96], so that it is often referred to as Nourein's method.

The Börsch-Supan's method (M_3) can be modified in the following way. Assuming that the quantity $t_i = \sum_{j \neq i} W_j / (z_i - z_j)$ is sufficiently small in magnitude, we use the approximation $1/(1 + t_i) \approx 1 - t_i$ and from (M_3) we obtain

Tanabe's method [88], [147], [171], order 3:

$$\hat{z}_i = z_i - W_i \left(1 - \sum_{\substack{j=1 \\ j \neq i}}^{n} \frac{W_j}{z_i - z_j} \right) \quad (i \in \mathbf{I}_n). \tag{M_8}$$

This method is often called Tanabe's method due to Tanabe [171] although it was known earlier (see M. Prešić [147], G. V. Milovanović [88]). Kanno, Kyurkchiev, and Yamamoto [69] have shown that Tanabe's method (M_8) may be obtained by applying Chebyshev's method

$$\hat{z} = z - \frac{f(z)}{f'(z)} \left(1 + \frac{f(z) f''(z)}{2 f'(z)^2} \right)$$

to the system (1.12).

Comment (M_4). The method (M_4) can be regarded as a modification of Ostrowski's method of the third order

$$\hat{z}_i = z_i - \frac{1}{\left[\delta_{1,i}^2 - \delta_{2,i} \right]^{1/2}} = z - \frac{1}{\left[\dfrac{P'(z_i)^2 - P(z_i) P''(z_i)}{P(z_i)^2} \right]^{1/2}}$$

(see Ostrowski [99], where the term *square root method* is used). The additional term in the form of the sum in (M_4) provides the simultaneous determination of all zeros of a polynomial and, at the same time, the increase of the order of convergence from 3 to 4.

It is interesting to note that the fixed point relation (F_4) was applied first for the construction of the simultaneous interval method in circular complex arithmetic, proposed by Gargantini [47]. The iterative formula (M_4) can be obtained as the approximation of the centers of resulting improved disks produced by Gargantini's interval method. The square root method (M_4) and its modifications have been considered in detail in [142].

Substituting ζ_j by Newton's approximation $z_j - N_j$ and Halley's approximation $z_j - H_j$ in (F_4), we obtain the following accelerated methods.

The square root method with Newton's corrections [142], order 5:

$$\hat{z}_i = z_i - \frac{1}{\left[\delta_{1,i}^2 - \delta_{2,i} - \displaystyle\sum_{\substack{j=1 \\ j \neq i}}^n \frac{1}{(z_i - z_j + N_j)^2}\right]^{1/2}} \quad (i \in \boldsymbol{I}_n). \tag{M_9}$$

The square root method with Halley's corrections [142], order 6:

$$\hat{z}_i = z_i - \frac{1}{\left[\delta_{1,i}^2 - \delta_{2,i} - \displaystyle\sum_{\substack{j=1 \\ j \neq i}}^n \frac{1}{(z_i - z_j + H_j)^2}\right]^{1/2}} \quad (i \in \boldsymbol{I}_n). \tag{M_{10}}$$

Both methods have very high computational efficiency since Newton's and Halley's corrections use the already calculated values $P(z_j)$, $P'(z_j)$, $P''(z_j)$ $(j \in \boldsymbol{I}_n)$.

Comment (M_5). As mentioned above, the fixed point relation (F_5) is a special case of a more general formula derived by X. Wang and Zheng in [182]. Consequently, the iterative method (M_5) is a special case of the family of iterative methods based on the generalized fixed point relation (1.16). This family has the order of convergence $k + 2$, where k is the highest order of derivatives of P appearing in the iterative formula. In special cases, for $k = 1$ and $k = 2$, one obtains the iterative methods (M_2) and (M_5), respectively. Modified methods which accelerate (M_5) have been considered in [143] and [177].

Wang–Zheng's method may also be considered as a method of Halley's type. Indeed, the function $f(z)$ given by (1.8) is the denominator of Halley's correction $H_i = 1/f(z_i)$ which appears in the well-known classical Halley's method $\hat{z}_i = z_i - H_i$ of the third order (see (1.10)).

Substituting ζ_j with Newton's approximation $z_j - N_j$ and Halley's approximation $z_j - H_j$ in (F_5), we obtain the following accelerated methods.

The Halley-like method with Newton's corrections [177], order 5:

$$\hat{z}_i = z_i - \frac{1}{f(z_i) - \dfrac{P(z_i)}{2P'(z_i)}\left[\left(\displaystyle\sum_{\substack{j=1 \\ j \neq i}}^n \frac{1}{z_i - z_j + N_j}\right)^2 + \displaystyle\sum_{\substack{j=1 \\ j \neq i}}^n \frac{1}{(z_i - z_j + N_j)^2}\right]}$$

$$(i \in \boldsymbol{I}_n). \tag{M_{11}}$$

The Halley-like method with Halley's corrections [177], order 6:

$$\hat{z}_i = z_i - \frac{1}{f(z_i) - \dfrac{P(z_i)}{2P'(z_i)}\left[\left(\displaystyle\sum_{\substack{j=1 \\ j \neq i}}^n \frac{1}{z_i - z_j + H_j}\right)^2 + \displaystyle\sum_{\substack{j=1 \\ j \neq i}}^n \frac{1}{(z_i - z_j + H_j)^2}\right]}$$

$$(i \in \boldsymbol{I}_n). \tag{M_{12}}$$

In [182] and [185], X. Wang and Zheng have derived a family of simultaneous methods based on Bell's polynomials (for details about Bell's polynomials, see [6], [154, Chap. 5]). To describe this family of simultaneous methods, we give an outline of Bell-like iterations.

For $z \in \mathbb{C}$, let us define

$$s_{\lambda,i} = \sum_{\substack{j=1 \\ j \neq i}}^{n} (z - \zeta_j)^{-\lambda} \tag{1.13}$$

and

$$\Delta_{0,i}(z) = 1, \quad \Delta_{k,i}(z) = \frac{(-1)^k}{k!} P(z) \frac{d^k}{dz^k} \left[P(z)^{-1} \right] \quad (k = 1, \ldots, n). \tag{1.14}$$

For $k \in \mathbb{N}$ and $(z_1, \ldots, z_k) \in \mathbb{C}^k$, the sum of the product of powers of $z_1, \ldots, z_k$

$$B_k(z_1, \ldots, z_k) := \sum_{\nu=1}^{k} \sum \prod_{\lambda=1}^{k} \frac{1}{q_\lambda!} \left(\frac{z_\lambda}{\lambda} \right)^{q_\lambda}, \quad B_0 = 1 \tag{1.15}$$

is called *Bell's polynomial*. The second sum on the right side runs over all nonnegative integers $q_1, \ldots, q_k$ satisfying the pair of equations

$$q_1 + 2q_2 + \cdots + kq_k = k,$$
$$q_1 + q_2 + \cdots + q_k = \nu \quad (1 \leq \nu \leq k).$$

For example, the first few B_i are

$$B_1(s_1) = s_1,$$
$$B_2(s_1, s_2) = \frac{1}{2}s_2 + \frac{1}{2}s_1^2,$$
$$B_3(s_1, s_2, s_3) = \frac{1}{3}s_3 + \frac{1}{2}s_2 s_1 + \frac{1}{6}s_1^3,$$

and so forth. We note that Bell's polynomials can be computed recurrently as

$$B_0 = 1, \quad B_k(z_1, \ldots, z_k) = \frac{1}{k} \sum_{\nu=1}^{k} z_\nu B_{k-\nu}(z_1, \ldots, z_k).$$

X. Wang and Zheng have derived in [182] the fixed point relation

$$\zeta_i = z - \frac{\Delta_{k-1,i}(z)}{\Delta_{k,i}(z) - B_k(s_{1,i}, \ldots, s_{k,i})} \quad (i \in \boldsymbol{I}_n), \tag{1.16}$$

where $s_{\lambda,i}$, $\Delta_{k,i}(z)$, and B_k are defined by (1.13), (1.14), and (1.15), respectively. This relation has effectively been applied to the construction of the following family of iterative methods for the simultaneous determination of all zeros of a polynomial putting $z = z_i$ and substituting ζ_j with z_j:

$$\hat{z}_i = z_i - \frac{\Delta_{k-1,i}(z_i)}{\Delta_{k,i}(z_i) - B_k(\hat{s}_{1,i}, \dots, \hat{s}_{k,i})} \quad (i \in \boldsymbol{I}_n),$$

where $\hat{s}_{\lambda,i} = \sum_{j \neq i}(z_i - z_j)^{-\lambda}$. The order of convergence of the basic method is $k + 2$, where k is the order of the highest derivative of polynomial used in the iterative formula. Two special cases of this family, which are obtained for $k = 1$ (the Ehrlich–Aberth's method (M_2)) and $k = 2$ (the Halley-like method (M_5)), are presented in this part.

Part II: Construction of Zero-Finding Methods by Weierstrass' Functions

As mentioned above, there are different procedures for constructing iterative methods for the simultaneous approximations of polynomial zeros; one of them, based on FPR, is described in Part I. It is of interest to apply the procedures that can generate a number of simultaneous methods in a unified way. Such a method, based on the application of Weierstrass' function, is presented in this part. A few of the presented results have already been given in [156]; we have extended here the mentioned approach to obtain other simultaneous methods.

WEIERSTRASS' FUNCTIONS

Our aim is to present a simple approach for the construction of iterative methods for the simultaneous determination of polynomial zeros. In this manner, it is possible to derive the numerous most frequently used simultaneous zero-finding methods, but in a simpler way than the original derivation. The proposed developing technique combines so-called Weierstrass' functions and suitable methods for finding a single (simple or multiple) zero of a function. Aside from the presented methods, this approach can also be applied to other specific methods for a single zero to develop known or new iterative formulae.

Let $P(z)$ be a monic polynomial of degree n with simple zeros $\zeta_1, \dots, \zeta_n$. Let us assume that $z_1, \dots, z_n$ are distinct points and define the polynomial $Q(z) = \prod_{j=1}^{n}(z - z_j)$. Then, using the development to partial fractions

$$\frac{P(z) - Q(z)}{Q(z)} = \sum_{j=1}^{n} \frac{a_j}{z - z_j}, \quad a_j = \frac{P(z) - Q(z)}{Q'(z)}\bigg|_{z=z_j} = \frac{P(z_j)}{\prod_{\substack{j=1 \\ j \neq i}}^{n}(z_i - z_j)} = W_j,$$

we derive

$$P(z) = \Big(\sum_{j=1}^{n} \frac{W_j}{z - z_j} + 1\Big) \prod_{j=1}^{n}(z - z_j) = \prod_{\substack{j=1 \\ j \neq i}}^{n}(z - z_j)\Big(W_i + (z - z_i)\Big(1 + \sum_{\substack{j=1 \\ j \neq i}}^{n} \frac{W_j}{z - z_j}\Big)\Big).$$

This representation of P can also be obtained using Lagrange's interpolation. Dividing the last relation by $\prod_{j \neq i}(z - z_j)$, we find

$$W_i(z) = \underbrace{\frac{P(z)}{\prod_{\substack{j=1 \\ j \neq i}}^{n}(z - z_j)}}_{F(1)} = W_i + \underbrace{(z - z_i)\Big(1 + \sum_{\substack{j=1 \\ j \neq i}}^{n} \frac{W_j}{z - z_j}\Big)}_{F(2)} \quad (i \in \boldsymbol{I}_n), \quad (1.17)$$

where we set $W_i(z_j) = W_j$. We have two equivalent forms $F(1)$ and $F(2)$ of the function $z \mapsto W_i(z)$ which will be called *Weierstrass' function*. The name comes from Weierstrass' iterative formula

$$\hat{z}_i = z_i - W_i \quad (i \in \boldsymbol{I}_n)$$

of the second order for the simultaneous determination of simple zeros of a polynomial P (see (M_1)).

In the following, we will use the abbreviations

$$G_{k,i} = \sum_{\substack{j=1 \\ j \neq i}}^{n} \frac{W_j}{(z_i - z_j)^k}, \quad S_{k,i} = \sum_{\substack{j=1 \\ j \neq i}}^{n} \frac{1}{(z_i - z_j)^k} \quad (k = 1, 2).$$

Let

$$N_i = \frac{P(z_i)}{P'(z_i)} = \frac{1}{\delta_{1,i}} \quad \text{and} \quad H_i = \frac{P(z_i)}{P'(z_i) - \dfrac{P(z_i)P''(z_i)}{2P'(z_i)}} = \frac{2\delta_{1,i}}{2\delta_{1,i}^2 - \delta_{2,i}}$$

$$(1.18)$$

be, respectively, Newton's and Halley's corrections appearing in the well-known iterative formulae

$$\hat{z}_i = z_i - N_i \quad \text{(Newton's method)},$$
$$\hat{z}_i = z_i - H_i \quad \text{(Halley's method)}.$$

The order of convergence of these methods is 2 and 3, respectively. Aside from approximations $z_1, \ldots, z_n$ to the zeros $\zeta_1, \ldots, \zeta_n$, we will use *improved* approximations c_j, where we take most frequently $c_j = z_j - N_j$ (Newton's approximation) or $c_j = z_j - H_j$ (Halley's approximation), see (1.18). Using the improved approximations c_j, we define the *modified Weierstrass' function*

$$\widetilde{W}_i(z) = \frac{P(z)}{\prod\limits_{\substack{j=1 \\ j \neq i}}^{n}(z - c_j)}. \tag{1.19}$$

Dealing with $\widetilde{W}_i(z)$, we denote the corresponding sum with $\widetilde{S}_{k,i} = \sum\limits_{\substack{j=1 \\ j \neq i}}^{n} \frac{1}{(z_i - c_j)^k}$ $(k = 1, 2)$.

The derivatives of $W_i(z)$ can be found using either (1.17)–$F(1)$ or (1.17)–$F(2)$. In the first case ($F(1)$), we use the logarithmic derivatives and find

$$\left.\frac{W_i'(z)}{W_i(z)}\right|_{z=z_i} = \delta_{1,i} - S_{1,i},$$

$$\left.\frac{W_i''(z)}{W_i'(z)}\right|_{z=z_i} = \delta_{1,i} - S_{1,i} + \frac{\delta_{2,i} - \delta_{1,i}^2 + S_{2,i}}{\delta_{1,i} - S_{1,i}}. \tag{1.20}$$

Similarly, using the logarithmic derivatives in (1.19), we obtain

$$\left.\frac{\widetilde{W}_i'(z)}{\widetilde{W}_i(z)}\right|_{z=z_i} = \delta_{1,i} - \widetilde{S}_{1,i},$$

$$\left.\frac{\widetilde{W}_i''(z)}{\widetilde{W}_i'(z)}\right|_{z=z_i} = \delta_{1,i} - \widetilde{S}_{1,i} + \frac{\delta_{2,i} - \delta_{1,i}^2 + \widetilde{S}_{2,i}}{\delta_{1,i} - \widetilde{S}_{1,i}}. \tag{1.21}$$

Starting from (1.17)–$F(2)$, it is easy to find

$$W_i'(z_i) = 1 + \sum_{\substack{j=1 \\ j \neq i}}^{n} \frac{W_j}{z_i - z_j} = 1 + G_{1,i},$$

$$W_i''(z_i) = -2 \sum_{\substack{j=1 \\ j \neq i}}^{n} \frac{W_j}{(z_i - z_j)^2} = -2G_{2,i}. \tag{1.22}$$

In this part, we demonstrate the application of Weierstrass' function and the modified Weierstrass' function in the construction of iterative methods for the simultaneous determination of polynomial zeros. Let us note that

Weierstrass' functions are rational functions whose nominator is a given polynomial P. All applications are based on the fact that the rational function W (or $\widetilde{W}$) has the same zeros as the polynomial P.

We emphasize that the use of corrections is justified only when their evaluations require the already calculated quantities. In this way, the order of convergence is increased using a negligible number of numerical operations, giving a high computational efficiency to the stated method. The most convenient corrections with the described property are Newton's approximation $c_i = z_i - N_i$ and Halley's approximation $c_i = z_i - H_i$ since they can be expressed using $\delta_{1,i}$ and $\delta_{2,i}$, see (1.18). The Weierstrass' approximation $c_j = z_i - W_i$ is suitable for the improvement on derivative-free methods, see, e.g., the simultaneous method (M_7).

The Börsch-Supan's method

As mentioned above, Weierstrass' function $z \mapsto W_i(z)$ has the same zeros $\zeta_1, \ldots, \zeta_n$ as the polynomial P. The Newton's method applied to Weierstrass' function, instead of $P(z)$, gives

$$\hat{z}_i = z_i - \frac{W_i(z_i)}{W_i'(z_i)}. \tag{1.23}$$

Substituting $W_i'(z_i)$ given by (1.22) into (1.23), we immediately obtain

$$\hat{z}_i = z_i - \frac{W_i}{1 + \sum_{\substack{j=1 \\ j \neq i}}^{n} \frac{W_j}{z_i - z_j}} = z_i - \frac{1}{1 + G_{1,i}} \quad (i \in \boldsymbol{I}_n), \tag{1.24}$$

which is cubically convergent Börsch-Supan's method presented in [10] (see (M_3)).

The Ehrlich–Aberth's method

Applying the logarithmic derivative to

$$W_i(z) = \frac{P(z)}{\prod_{\substack{j=1 \\ j \neq i}}^{n} (z - z_j)},$$

we find

$$\frac{W_i'(z_i)}{W_i(z_i)} = \frac{P'(z_i)}{P(z_i)} - \sum_{\substack{j=1 \\ j \neq i}}^{n} \frac{1}{z_i - z_j},$$

so that (1.23) yields

$$\hat{z}_i = z_i - \cfrac{1}{\cfrac{P'(z_i)}{P(z_i)} - \sum_{\substack{j=1 \\ j \neq i}}^{n} \cfrac{1}{z_i - z_j}} \quad (i \in \boldsymbol{I}_n).$$

This is the well-known iterative method of the third order considered in [1], [9], [33], [85], frequently called the Ehrlich–Aberth's method (see (M_2)).

If we apply the modified Weierstrass' function (with $c_j = z_j - N_j$) in the same manner, we obtain

$$\frac{\widetilde{W}_i'(z_i)}{\widetilde{W}_i(z_i)} = \frac{P'(z_i)}{P(z_i)} - \sum_{\substack{j=1 \\ j \neq i}}^{n} \frac{1}{z_i - z_j + N_j}.$$

From (1.23), it follows

$$\hat{z}_i = z_i - \cfrac{1}{\cfrac{P'(z_i)}{P(z_i)} - \sum_{\substack{j=1 \\ j \neq i}}^{n} \cfrac{1}{z_i - z_j + N_j}} \quad (i \in \boldsymbol{I}_n),$$

which is the fourth-order Ehrlich–Aberth's method with Newton's corrections, also known as Nourein's method [97] (see (M_6)).

The Ostrowski-like methods

Applying the Ostrowski's method [99]

$$\hat{z} = z - \frac{P(z)}{\sqrt{P'(z)^2 - P(z)P''(z)}} \tag{1.25}$$

to Weierstrass' function $W_i(z)$ instead of $P(z)$, and using the derivatives given by (1.22), we obtain

$$\hat{z}_i = z_i - \frac{W_i}{\sqrt{(1 + G_{1,i})^2 + 2W_i G_{2,i}}} \quad (i \in \boldsymbol{I}_n).$$

This fourth-order method was also derived in [124] using Hansen–Patrick's family.

Let us rewrite (1.25) in the form

$$\hat{z} = z - \cfrac{1}{\left[\left(\cfrac{P'(z)}{P(z)} \right)^2 - \cfrac{P''(z)}{P'(z)} \cdot \cfrac{P'(z)}{P(z)} \right]^{1/2}}$$

and substitute the quotients $P'(z_i)/P(z_i)$ and $P''(z_i)/P'(z_i)$ by $W_i'(z_i)/W_i(z_i)$ and $W_i''(z_i)/W_i'(z_i)$ (given by (1.20)) and then by $\widetilde{W}_i'(z_i)/\widetilde{W}_i(z_i)$ and $\widetilde{W}_i''(z_i)/\widetilde{W}_i'(z_i)$ (given by (1.21)). Thus, we obtain the following iterative methods for the simultaneous determination of polynomial zeros

$$\hat{z}_i = z_i - \cfrac{1}{\left[\delta_{1,i}^2 - \delta_{2,i} - \displaystyle\sum_{\substack{j=1 \\ j \neq i}}^{n} \frac{1}{(z_i - z_j)^2} \right]^{1/2}} \quad (i \in \boldsymbol{I}_n), \qquad (1.26)$$

$$\hat{z}_i = z_i - \cfrac{1}{\left[\delta_{1,i}^2 - \delta_{2,i} - \displaystyle\sum_{\substack{j=1 \\ j \neq i}}^{n} \frac{1}{(z_i - z_j + N_j)^2} \right]^{1/2}} \quad (i \in \boldsymbol{I}_n), \qquad (1.27)$$

$$\hat{z}_i = z_i - \cfrac{1}{\left[\delta_{1,i}^2 - \delta_{2,i} - \displaystyle\sum_{\substack{j=1 \\ j \neq i}}^{n} \frac{1}{(z_i - z_j + H_j)^2} \right]^{1/2}} \quad (i \in \boldsymbol{I}_n), \qquad (1.28)$$

derived in a different way in [142]. The order of convergence of the methods (1.26), (1.27), and (1.28) is 4, 5, and 6, respectively (see (M_4), (M_9), and (M_{10})).

BILINEAR FUNCTION AND TWO FOURTH-ORDER METHODS

We approximate the Weierstrass' function $W_i(z)$ at the point $z = z_i$ by the *bilinear function* g of the form

$$g(z) = \frac{(z - z_i) + \alpha_1}{\alpha_2(z - z_i) + \alpha_3} \quad (z_i, \alpha_1, \alpha_2, \alpha_3 \in \mathbb{C}), \qquad (1.29)$$

which coincides with $W_i(z)$ at z_i up through second derivatives, i.e.,

$$g^{(k)}(z_i) = W_i^{(k)}(z_i) \quad (k = 0, 1, 2, \ W_i^{(0)}(z) \equiv W_i(z)). \qquad (1.30)$$

Let $\hat{z}_i$ be a complex number such that $g(\hat{z}_i) = 0x$. Then from (1.29), we obtain

$$\hat{z}_i = z_i - \alpha_1. \qquad (1.31)$$

This means that if z_i is a sufficiently good approximation to a zero of the rational function $W_i(z)$ (and, thus, a zero of the polynomial P), then $\hat{z}_i$ is an improved approximation to that zero.

To find the unknown complex coefficient α_1, we start from (1.30) and get the system of equations

$$\frac{\alpha_1}{\alpha_3} = W_i(z_i), \quad \frac{\alpha_3 - \alpha_1\alpha_2}{\alpha_3^2} = W_i'(z_i), \quad \frac{2\alpha_2(\alpha_2\alpha_1 - \alpha_3)}{\alpha_3^2} = W_i''(z_i).$$

Hence

$$\alpha_1 = \frac{2W_i(z_i)W_i'(z_i)}{2W_i'(z_i)^2 - W_i(z_i)W_i''(z_i)}$$

and, according to (1.31), we find

$$\hat{z}_i = z_i - \frac{2W_i(z_i)W_i'(z_i)}{2W_i'(z_i)^2 - W_i(z_i)W_i''(z_i)} \quad (i \in \boldsymbol{I}_n). \tag{1.32}$$

The derivatives of $W_i(z)$ can be found using either (1.17)–$F(1)$ or (1.17)–$F(2)$. In the first case ($F(1)$), we use (1.20) and construct from (1.32) the following fourth-order iterative method for the simultaneous approximation of all simple zeros of a polynomial P:

$$\hat{z}_i = z_i - \frac{2(S_{1,i} - \delta_{1,i})}{\delta_{2,i} - 2\delta_{1,i}^2 + 2S_{1,i}\delta_{1,i} + S_{2,i} - S_{1,i}^2} \quad (i \in \boldsymbol{I}_n).$$

This iterative formula was derived (in a completely different way) by Sakurai, Torri, and Sugiura in [157].

When we use the derivatives of $W_i(z)$ at $z = z_i$ given by (1.22), from (1.32), we find

$$\hat{z}_i = z_i - \frac{W_i(1 + G_{1,i})}{(1 + G_{1,i})^2 + W_i G_{2,i}} \quad (i \in \boldsymbol{I}_n). \tag{1.33}$$

The method (1.33) was derived by Ellis and Watson [34] using an entirely different approach.

The Börsch-Supan's method with Weierstrass' correction

Let $i \in \boldsymbol{I}_n$ be fixed and let $z_1, \ldots, z_n \in \mathbb{C}$. Using two approximations z' and z'' for ζ_i, $z' \neq z''$, the new approximation $\hat{z}$ obtained by the secant method applied to $W_i(z)$ is

$$\hat{z} := z' - \frac{z'' - z'}{W_i(z'') - W_i(z')} W_i(z'). \tag{1.34}$$

Clearly, we assume that $W_i(z'') \neq W_i(z')$.

Let $z' = z_i$ and $z'' = z_i - W_i$. Then by (1.17)–$F(2)$, we have

$$\frac{W_i(z')}{W_i(z'') - W_i(z')} = -\Big(1 + \sum_{\substack{j=1 \\ j \neq i}}^{n} \frac{W_j}{z_i - W_i - z_j}\Big)^{-1}.$$

According to this and (1.34), we obtain

$$
\hat{z} = z' - (z'' - z')\frac{W_i(z')}{W_i(z'') - W_i(z')}
$$

$$
= z_i - \frac{W_i}{1 + \displaystyle\sum_{\substack{j=1 \\ j \neq i}}^{n} \frac{W_j}{z_i - W_i - z_j}} \qquad (i \in \boldsymbol{I}_n). \tag{1.35}
$$

The iterative method (1.35) was derived by Nourein in [96] in a different way and has the order of convergence equal to 4. Let us note that (1.35) is the improvement of the method (1.24) (see (M_3) and (M_7)).

The iterative method (1.35) can also be derived starting from the Steffensen-like iterative formula (see Steffensen [169] and Ostrowski [99, p. 245])

$$
\hat{z}_i = z_i - \frac{P(z_i)^2}{P(z_i) - P(z_i - P(z_i))} \tag{1.36}
$$

taking Weierstrass' function $W_i(z)$ instead of the polynomial $P(z)$ in (1.36). Using (1.17)–$F(2)$ and (1.36), we then obtain

$$
\hat{z}_i = z_i - \frac{W_i(z_i)^2}{W_i(z_i) - W_i(z_i - W_i(z_i))}
$$

$$
= z_i - \frac{W_i^2}{W_i - \left[W_i - W_i\left(1 + \displaystyle\sum_{\substack{j=1 \\ j \neq i}}^{n} \frac{W_j}{z_i - W_i - z_j}\right)\right]}
$$

$$
= z_i - \frac{W_i}{1 + \displaystyle\sum_{\substack{j=1 \\ j \neq i}}^{n} \frac{W_j}{z_i - W_i - z_j}} \qquad (i \in \boldsymbol{I}_n).
$$

The Zheng–Sun's method

We start from the iterative formula (1.35) and use the development into geometric series (assuming that $|W_i|$ is sufficiently small) to obtain

$$
\hat{z} = z_i - \frac{W_i}{1 + \displaystyle\sum_{\substack{j=1 \\ j \neq i}}^{n} \frac{W_j}{z_i - W_i - z_j}} = z_i - \frac{W_i}{1 + \displaystyle\sum_{\substack{j=1 \\ j \neq i}}^{n} \frac{W_j}{(z_i - z_j)\left(1 - \dfrac{W_i}{z_i - z_j}\right)}}
$$

$$
= z_i - \frac{W_i}{1 + \displaystyle\sum_{\substack{j=1 \\ j \neq i}}^{n} \frac{W_j}{(z_i - z_j)}\left(1 + \dfrac{W_i}{z_i - z_j} + \left(\dfrac{W_i}{z_i - z_j}\right)^2 + \cdots\right)}.
$$

Neglecting the terms of higher order, from the last relation, we get

$$\hat{z} = z_i - \cfrac{W_i}{1 + \sum_{\substack{j=1 \\ j \neq i}}^{n} \cfrac{W_j}{z_i - z_j} + W_i \sum_{\substack{j=1 \\ j \neq i}}^{n} \cfrac{W_j}{z_i - z_j}} = z_i - \cfrac{W_i}{1 + G_{1,i} + W_i G_{2,i}} \quad (i \in I_n).$$

The last iterative formula defines the iterative method of the fourth order proposed by Zheng and Sun in [196]. Let us note that their derivation of this method is unnecessarily complicated and occupies almost three pages in [196].

The Chebyshev-like methods

Applying the Chebyshev's third-order method [172]

$$\hat{z} = z - \frac{P(z)}{P'(z)} \left(1 + \frac{P(z)}{P'(z)} \cdot \frac{P''(z)}{2P'(z)} \right)$$

to Weierstrass' function $W_i(z)$, we obtain

$$\hat{z}_i = z_i - \frac{W_i}{W_i'} \left(1 + \frac{W_i}{W_i'} \cdot \frac{W_i''}{2W_i'} \right) \quad (i \in I_n). \tag{1.37}$$

Hence, by (1.20), we generate the fourth-order iterative method

$$\hat{z}_i = z_i - \frac{1}{2(\delta_{1,i} - S_{1,i})} \left[3 + \frac{\delta_{2,i} - \delta_{1,i}^2 + S_{2,i}}{(\delta_{1,i} - S_{1,i})^2} \right] \quad (i \in I_n), \tag{1.38}$$

proposed by Sakurai and Petković [156].

Similarly, applying (1.22)–(1.37), we obtain another fourth-order method of Chebyshev's type

$$\hat{z}_i = z_i - \frac{W_i}{1 + G_{1,i}} \left(1 - \frac{W_i G_{2,i}}{(1 + G_{1,i})^2} \right) \quad (i \in I_n),$$

considered in [146].

If we use the modified Weierstrass' function given by (1.19) and the corresponding expressions (1.21), then we obtain the accelerated Chebyshev-like method

$$\hat{z}_i = z_i - \frac{1}{2(\delta_{1,i} - \widetilde{S}_{1,i})} \left[3 + \frac{\delta_{2,i} - \delta_{1,i}^2 + \widetilde{S}_{2,i}}{(\delta_{1,i} - \widetilde{S}_{1,i})^2} \right] \quad (i \in I_n). \tag{1.39}$$

It is not difficult to prove that the order of convergence of the method (1.39) is 5 if we employ Newton's approximations $c_j = z_j - N_j$, and 6 when Halley's approximations $c_j = z_j - H_j$ are applied.

The cubic derivative-free method

Let us take the polynomial $P(z)$ instead of $W_i(z)$ in (1.34). With the same approximations, i.e., $z' = z_i$, $z'' = z_i - W_i$, from (1.34), we obtain the cubically convergent derivative-free method

$$\hat{z}_i = z_i - \frac{W_i}{1 - \dfrac{P(z_i - W_i)}{P(z_i)}} \quad (i \in \boldsymbol{I}_n).$$

To avoid confusion, we emphasize that $z_i - W_i$ is the argument of P in the last iterative formula.

METHODS FOR MULTIPLE ZEROS

Let us consider a polynomial P with multiple zeros $\zeta_1, \ldots, \zeta_\nu$ of the known multiplicities $\mu_1, \ldots, \mu_\nu$ ($\mu_1 + \cdots + \mu_\nu = n$) and let $c_1, \ldots, c_\nu$ be some approximations to these zeros. In a similar way to the previous part, we can construct methods for the simultaneous determination of multiple zeros using the Weierstrass-like function

$$W_i^*(z) = P(z) / \prod_{\substack{j=1 \\ j \neq i}}^{\nu} (z - c_j)^{\mu_j} \tag{1.40}$$

and its derivatives in suitable iterative formulae for finding a single multiple zero of a function f (not necessary a polynomial) of the multiplicity μ. For example, we give several existing formulae of this kind, assuming that the considered function is an algebraic polynomial P:

$$\hat{z} = z - \mu \frac{P(z)}{P'(z)} \quad \text{(Schröder's method [160])}, \tag{1.41}$$

$$\hat{z} = z - \frac{2}{\dfrac{\mu+1}{\mu} \dfrac{P'(z)}{P(z)} - \dfrac{P''(z)}{P'(z)}} \quad \text{(Halley-like method [55])}, \tag{1.42}$$

$$\hat{z} = z - \frac{\sqrt{\mu} P(z)}{\sqrt{P'(z)^2 - P(z) P''(z)}} \quad \text{(square root method [99])}, \tag{1.43}$$

$$\hat{z} = z - \frac{nP(z)}{P'(z) \pm \sqrt{\dfrac{n-\mu}{\mu} \left[(n-1)P'(z)^2 - nP(z)P''(z) \right]}} \quad \text{(see [55])}, \tag{1.44}$$

$$\hat{z} = z - \frac{\mu(\alpha+1)P(z)}{\alpha P'(z) \pm \sqrt{(\mu(\alpha+1) - \alpha)P'(z)^2 - \mu(\alpha+1)P(z)P''(z)}}, \tag{1.45}$$

etc. The formula (1.45) was derived in [135] and it is a simplified version of Hansen–Patrick's formula (4.7). The above formulae are of the form $\hat{z}_i = z_i - c_i$; c_i is just the correction appearing in (1.40) (taking $z = z_i$ and $\mu = \mu_i$

in (1.41)–(1.45)). The iterative method (1.41) has quadratic convergence; methods (1.42)–(1.45) possess cubic convergence.

Starting from (1.40) and associating the sum

$$S_{k,i}^* = \sum_{\substack{j=1 \\ j\neq i}}^{\nu} \frac{\mu_j}{(z_i - c_j)^k} \quad (k = 1, 2)$$

to the Weierstrass-like function (1.40), by logarithmic derivatives, we find

$$
\begin{aligned}
\left. \frac{\left(W_i^*(z)\right)'}{W_i^*(z)} \right|_{z=z_i} &= \delta_{1,i} - S_{1,i}^*, \\
\left. \frac{\left(W_i^*(z)\right)''}{\left(W_i^*(z)\right)'} \right|_{z=z_i} &= \delta_{1,i} - S_{1,i}^* + \frac{\delta_{2,i} - \delta_{1,i}^2 + S_{2,i}^*}{\delta_{1,i} - S_{1,i}^*}.
\end{aligned}
\tag{1.46}
$$

For illustration, we will consider the iterative formulae (1.41) and (1.42). Substituting $P'(z_i)/P(z_i)$ with $(W_i^*(z_i))'/W_i(z_i)$ (given by (1.46)) in (1.41) and taking $c_j = z_j$ and $c_j = z_j - \mu_j/\delta_{1,j}$, we obtain, respectively, the third-order method

$$\hat{z}_i = z_i - \frac{\mu_i}{\delta_{1,i} - \sum_{\substack{j=1 \\ j\neq i}}^{\nu} \dfrac{\mu_j}{z_i - z_j}} \quad (i \in \boldsymbol{I}_\nu)$$

and the fourth-order method

$$\hat{z}_i = z_i - \frac{\mu_i}{\delta_{1,i} - \sum_{\substack{j=1 \\ j\neq i}}^{\nu} \dfrac{\mu_j}{z_i - z_j + \mu_j/\delta_{1,j}}} \quad (i \in \boldsymbol{I}_\nu)$$

for the simultaneous approximation of all multiple zeros of the polynomial P.

Substituting $P'(z_i)/P(z_i)$ with $(W_i^*(z_i))'/W_i(z_i)$ and $P''(z_i)/P'(z_i)$ with $(W_i^*(z_i))''/(W_i^*(z_i))'$ in (1.42), using (1.46), we find

$$\hat{z}_i = z_i - \frac{2\mu_i(\delta_{1,i} - S_{1,i}^*)}{(\delta_{1,i} - S_{1,i}^*)^2 - \mu_i(\delta_{2,i} - \delta_{1,i}^2 + S_{2,i}^*)} \quad (i \in \boldsymbol{I}_\nu). \tag{1.47}$$

Taking $c_j = z_j$ in the sum $S_{1,i}^*$, the iterative formula (1.47) yields the fourth-order method for multiple zeros. The acceleration of the convergence can be achieved taking $c_j = z_j - N_j^*$ (the order 5) and $c_j = z_j - H_j^*$ (the order 6), where

$$N_j^* = \mu_j \frac{P(z_j)}{P'(z_j)} = \frac{\mu_j}{\delta_{1,j}}$$

and

$$H_j^* = \frac{1}{\dfrac{\mu_j + 1}{2\mu_j} \cdot \dfrac{P'(z_j)}{P(z_j)} - \dfrac{P''(z_j)}{2P'(z_j)}} = \frac{2\mu_j \delta_{1,j}}{(\mu_j + 1)\delta_{1,j}^2 - \mu_j \delta_{2,j}}$$

are the Schröder's correction and the Halley-like correction, respectively (see (1.41) and (1.42)).

In a similar way, the iterative formulae (1.43)–(1.45) can be modified to generate iterative methods for the simultaneous determination of multiple zeros, some of which are new ones.

All presented simultaneous methods are based on iterative methods for finding a single (simple or multiple) zero and derived in a simpler way compared with the original derivations. We can continue to apply other iterative methods to the Weierstrass' function $W_i(z)$, the modified Weierstrass' function $\widetilde{W}(z)$, and the Weierstrass-like function $W_i^*(z)$ to construct (existing or new) classes of simultaneous methods for finding polynomial zeros.

Part III: Approximations of Derivatives

Farmer and Loizou [37] showed that the substitution of approximations of a given polynomial

$$P(z) = a_n z^n + a_{n-1} z^{n-1} + \cdots + a_1 z + a_0, \quad a_n a_0 \neq 0$$

and its derivatives in iterative formulae for finding a single zero of a function can lead to a new class of methods for the simultaneous determination of polynomial zeros.

Let us define

$$u_i = z_i - \zeta_i, \quad A_k(z) = \frac{P^{(k)}(z)}{k! P'(z)},$$

$$S_k(z_i) = \sum_{j \neq i} \frac{1}{(z_i - z_j)^k}, \quad T_k(z_i) = \sum_{j \neq i} \frac{1}{(z_i - \zeta_j)^k} \quad (k = 1, 2, \ldots).$$

Farmer and Loizou [37] gave the following formulae:

$$P'(z_i) = a_n \prod_{j \neq i} (z_i - z_j) + \mathcal{O}(u), \tag{1.48}$$

$$A_2(z_i) = S_1(z_i) + \mathcal{O}(u), \tag{1.49}$$

$$2A_3(z_i) = A_2^2(z_i) - S_2(z_i) + \mathcal{O}(u), \tag{1.50}$$

$$3A_4(z_i) = 3A_2(z_i)A_3(z_i) - A_2^3(z_i) + S_3(z_i) + \mathcal{O}(u), \tag{1.51}$$

etc., where $u = \max_{1 \leq i \leq n} |u_i|$. The derivation of these formulae is elementary but cumbersome. For demonstration, we will prove (1.49).

Using the logarithmic differentiation, we start from the factorization

$$P(z) = a_n \prod_{j=1}^{n} (z - \zeta_j)$$

and obtain

$$\frac{P'(z)}{P(z)} = \sum_{j=1}^{n} \frac{1}{z - \zeta_j}, \tag{1.52}$$

wherefrom

$$P'(z) = P(z) \sum_{j=1}^{n} \frac{1}{z - \zeta_j}. \tag{1.53}$$

By differentiating (1.53), we find

$$P''(z) = P'(z) \sum_{j=1}^{n} \frac{1}{z - \zeta_j} - P(z) \sum_{j=1}^{n} \frac{1}{(z - \zeta_j)^2}. \tag{1.54}$$

Using (1.52), from (1.54), we obtain for $z = z_i$

$$\frac{P''(z_i)}{P'(z_i)} = \sum_{j=1}^{n} \frac{1}{z_i - \zeta_j} - \frac{1}{\sum_{j=1}^{n} (z_i - \zeta_j)^{-1}} \sum_{j=1}^{n} \frac{1}{(z_i - \zeta_j)^2}$$

$$= \frac{1}{u_i}(1 + u_i T_1(z_i)) - \frac{1}{u_i}\left(\frac{1 + u_i^2 T_2(z_i)}{1 + u_i T_1(z_i)}\right).$$

Developing $1/(1 + u_i T_1(z_i))$ into geometric series, we get

$$2A_2(z_i) = \frac{P''(z_i)}{P'(z_i)}$$

$$= \frac{1}{u_i}(1 + u_i T_1(z_i)) - \frac{1}{u_i}(1 + u_i^2 T_2(z_i))(1 - u_i T_1(z_i) + \mathcal{O}(u))$$

$$= 2T_1(z_i) + \mathcal{O}(u) = 2 \sum_{j \neq i} \frac{1}{(z_i - z_j)(1 + u_j/(z_i - z_j))} + \mathcal{O}(u)$$

$$= 2S_1(z_i) + \mathcal{O}(u),$$

where we have used again the development into geometric series. Therefore, we have proved (1.49).

Approximations (1.48)–(1.51) can be suitably applied to iterative processes for finding a single zero. In what follows, we will consider a class of methods of arbitrary order of convergence presented, e.g., by Farmer and Loizou [37]. Let $h(z) = P(z)/P'(z)$ and define λ_n recursively in the following way:

$$\lambda_1(z) = h(z),$$

$$\lambda_k(z) = \frac{h(z)}{1 - \sum\limits_{i=2}^{k} A_i(z) \prod\limits_{p=k-i+1}^{k-1} \lambda_p(z)} \quad (k = 2, 3, \ldots, n). \quad (1.55)$$

Then, the iterative method

$$\Psi_n \equiv \hat{z}_i = z_i - \lambda_n(z_i) \quad (m = 0, 1, \ldots) \quad (1.56)$$

is of order $n + 1$.

Remark 1.1. It is interesting to note that M. Petković and D. Herceg have shown in [113] that the class of methods (1.55) is equivalent to other classes of iterative methods, derived in various ways and expressed in different forms by Gerlach [51], Ford and Pennline [41], Wang [179], Varjuhin and Kasjan-juk [176], Jovanović [68], and Igarashi and Nagasaka [63].

The first few Ψ_n are given by

$$\Psi_3 = z_i - \frac{h(z_i)}{1 - A_2(z_i)h(z_i)} \quad \text{(Halley [54])},$$

$$\Psi_4 = z_i - \frac{h(z_i)\big[1 - A_2(z_i)h(z_i)\big]}{1 - 2A_2(z_i)h(z_i) + A_3(z_i)h(z_i)^2},$$

$$\Psi_5 = z_i - \frac{h(z_i)\big[1 - 2A_2(z_i)h(z_i) + A_3(z_i)h(z_i)^2\big]}{1 - 3A_2(z_i)h(z_i) + \big[2A_3(z_i) + A_2^2(z_i)\big]h(z_i)^2 - A_4(z_i)h(z_i)^3},$$

and so forth.

By replacing the derivative of the highest order in Ψ_n, and using an approximation of the type given by (1.48)–(1.51), we obtain iterative methods for the simultaneous determination of polynomial zeros.

The second-order iterative method (Weierstrass' or Durand–Kerner's method):

$$\hat{z}_i = z_i - \frac{P(z_i)}{a_n \prod\limits_{\substack{j=1 \\ j \neq i}}^{n} (z_i - z_j)} \quad (i \in I_n).$$

The third-order iterative method (Ehrlich–Aberth's method):

$$\hat{z}_i = z_i - \frac{h(z_i)}{1 - h(z_i)S_1(z_i)} \quad (i \in I_n).$$

The fourth-order iterative method (Farmer–Loizou's method):

$$\hat{z}_i = z_i - \frac{h(z_i)\big[1 - A_2(z_i)h(z_i)\big]}{1 - 2A_2(z_i)h(z_i) + \frac{1}{2}\big[A_2^2(z_i) - S_2(z_i)\big]h(z_i)^2} \quad (i \in \boldsymbol{I}_n),$$

etc.

1.2 Localization of Polynomial Zeros

Before applying any iterative method for the inclusion of polynomial zeros, it is necessary to find initial regions (disks or rectangles) containing these zeros. Obviously, these inclusion disks can be suitably used for iterative methods realized in ordinary ("point") real or complex arithmetic (for example, taking the centers of these complex intervals). There are a lot of results concerning this topic, from the classical ones presented in Henrici [57] and Marden [87] to the more recent contributions.

The choice of initial inclusion regions which contain polynomial zeros is strongly connected with the conditions for the convergence of iterative methods. Most of these conditions presented in the literature depend on unknown data, for instance, of some functions of the sought zeros, which is not of practical importance. In this section, we consider initial regions which depend on the initial complex approximations $z_1^{(0)}, \ldots, z_n^{(0)}$ to the simple complex zeros $\zeta_1, \ldots, \zeta_n$.

We begin with a particular result which has a global character. Consider a monic polynomial P of degree n,

$$P(z) = z^n + a_{n-1}z^{n-1} + \cdots + a_1 z + a_0 = \prod_{j=1}^{n}(z - \zeta_j), \quad (a_i \in \mathbb{C}), \quad (1.57)$$

with the zeros $\zeta_1, \ldots, \zeta_n$. Solving polynomial equations, it is often of interest to find an *inclusion radius* R for the given polynomial P such that all zeros of P satisfy

$$|\zeta_i| \le R \quad (i = 1, \ldots, n).$$

The following assertion has a great practical importance (see, e.g., [57, p. 457]).

Theorem 1.1. *Let* $\omega_1, \ldots, \omega_n$ *be positive numbers such that* $\omega_1 + \cdots + \omega_n \le 1$ *and let*

$$R := \max_{1 \le k \le n} \omega_k^{-1/k} |a_{n-k}|^{1/k}.$$

Then R is an inclusion radius for P.

Specially, taking $\omega_k = 1/2^k$, from Theorem 1.1, it follows that the disk centered at the origin with the radius

$$R = 2 \max_{1 \leq k \leq n} |a_{n-k}|^{1/k} \tag{1.58}$$

contains all zeros of the polynomial P. The last result can also be found in Dekker's work [28] and Knuth's book [76]. Note that it is often convenient in a subdividing procedure to take the smallest possible *square* containing the circle $\{z : |z| \leq R\}$ as the initial region, where R is given by (1.58).

Using (1.58) and the substitution $w = 1/z$, it is easy to show that the disk centered at the origin with the radius

$$r = \frac{1}{2} \min_{1 \leq k \leq n} \left| \frac{a_0}{a_k} \right|^{1/k}$$

does not contain any zero of P. Therefore, all zeros of the polynomial P lie in the annulus $\{z : r \leq |z| \leq R\}$.

The inclusion disk given by (1.58) has the center at the origin. But it is reasonable to translate this center at the center of gravity of the zeros $\zeta_1, \ldots, \zeta_n$. Since $\zeta_1 + \cdots + \zeta_n = -a_{n-1}$, the center of gravity is given by $c = -a_{n-1}/n$. Substituting $y = z + c = z - a_{n-1}/n$ in (1.57) transforms the polynomial P into the shifted polynomial

$$P(z + c) = z^n + b_{n-2}z^{n-2} + \cdots + b_1 z + b_0 \quad (b_{n-1} = 0).$$

The center of gravity of the zeros $\xi_1, \ldots, \xi_n$ ($\xi_i = \zeta_i + c$) of the transformed polynomial $P(z + c)$ is 0. The inclusion radius

$$R' = 2 \max_{2 \leq k \leq n} |b_{n-k}|^{1/k}$$

for the zeros $\xi_1, \ldots, \xi_n$ is most frequently less than R calculated by (1.58).

The following result of Van der Sluis [175] precisely describes the aforementioned procedure.

Theorem 1.2. *All the zeros* $\zeta_1, \ldots, \zeta_n$ *of the polynomial* (1.57) *satisfy*

$$\sqrt{\frac{2}{n}} L \leq |\zeta_i - c| \leq \frac{1 + \sqrt{5}}{2} L < 1.62 \, L,$$

where $c = -a_{n-1}/n$, $L = \max_{k \geq 2} |b_{n-k}|^{1/k}$, *and* $P(z + c) = \sum_{k=0}^{n} b_k z^k$ $(b_{n-1} = 0)$.

We have already derived the Durand–Kerner's iterative formula (see (M_1))

$$\hat{z}_i = z_i - \frac{P(z_i)}{\displaystyle\prod_{\substack{j=1 \\ j \neq i}}^{n} (z_i - z_j)} = z_i - W_i \quad (i \in \boldsymbol{I}_n). \tag{1.59}$$

Weierstrass' corrections W_i have been often used as a posteriori error estimates for a given set of approximate zeros. Smith [168] showed that the disk

$$|z - (z_i - W_i)| \leq (n-1)|W_i|$$

contains at least one zero of P. This is a slight improvement of the result of Braess and Hadeler [11] who proved that the disk given by

$$|z - z_i| \leq n|W_i|$$

also contains at least one zero of the polynomial P. The latter is a consequence of Smith's result.

Throughout this book, a disk with center c and radius r will be denoted by the parametric notation $\{c; r\}$. It is easy to show by the continuity that, if disks $\{z_1; n|W_1|\}, \ldots, \{z_n; n|W_n|\}$ are mutually disjoint, then each of them contains one and only one zero of P. The same is valid for Smith's disks.

We now present inclusion disks based on Weierstrass' corrections. It is known (see Carstensen [14] or Elsner [35]) that the characteristic polynomial of the $n \times n$-matrix

$$B := \mathrm{diag}\ (z_1, \ldots, z_n) - \begin{bmatrix} 1 \\ \vdots \\ 1 \end{bmatrix} \cdot [W_1 \cdots W_n]$$

is equal to $(-1)^n P(z)$. Hence, by Gerschgorin's inclusion theorem applied to B, we can get locations of the zeros of P. Before doing this, we may transform the matrix B into $T^{-1}BT$ having the same eigenvalues for any regular matrix T. The question "which T gives the best inclusion disks?" can be answered (in some sense) if T belongs to the class of diagonal matrices. It turns out that the "best Gerschgorin's disks" lead to the following estimate, proved by Carstensen [14] and Elsner [35].

Theorem 1.3. *For $p \in \{1, 2, \ldots, n\}$ and $\xi \in \mathbb{C}$, let r be a positive number bounded by*

$$\max_{1 \leq j \leq p} (|z_j - W_j - \xi| - |W_j|) < r < \min_{p+1 \leq j \leq n} (|z_j - W_j - \xi| + |W_j|)$$

such that $1 > y(r) \geq 0$, where

$$y(r) := \sum_{j=1}^{p} \frac{|W_j|}{r - |z_j - W_j - \xi| + |W_j|} + \sum_{j=p+1}^{n} \frac{|W_j|}{|z_j - W_j - \xi| + |W_j| - r}.$$

Then, there are exactly p zeros in the open disk with center ξ and radius r.

Remark 1.2. In the case $p = n$, the conditions on the upper bound of r and the last sum must be neglected. A reordering leads to more flexibility in Theorem 1.3.

Remark 1.3. Adopting notations from Theorem 1.3, it follows from $y(r) \leq 1$ by continuity that at least p zeros of P lie in the closed disk with center ξ and radius r.

In the case $p = 1$, Theorem 1.3 can be specified giving the following simpler estimate proved by Carstensen [13, Satz 3] (see also [9], [11], [14] for similar results).

Theorem 1.4. *Let $\eta_i := z_i - W_i \in \mathbb{C} \setminus \{z_1, \ldots, z_n\}$ and set*

$$\gamma_i := |W_i| \cdot \max_{\substack{1 \leq j \leq n \\ j \neq i}} |z_j - \eta_i|^{-1}, \quad \sigma_i := \sum_{\substack{j=1 \\ j \neq i}}^{n} \frac{|W_j|}{|z_j - \eta_i|}, \quad i \in \{1, \ldots, n\}.$$

If $\sqrt{1 + \gamma_i} > \sqrt{\gamma_i} + \sqrt{\sigma_i}$, then there is exactly one zero of P in the disk with center η_i and radius

$$|W_i| \cdot \left(1 - \frac{2(1 - 2\sigma_i - \gamma_i)}{1 - \sigma_i - 2\gamma_i + \sqrt{(1 - \sigma_i - 2\gamma_i)^2 + 4\sigma_i(1 - 2\sigma_i - \gamma_i)^2)}}\right).$$

If

$$\sqrt{1 + \gamma_i} > \sqrt{\gamma_i} + \sqrt{\sigma_i} \quad \text{and} \quad \gamma_i + 2\sigma_i < 1, \tag{1.60}$$

then there is exactly one zero of P in the disk with center η_i and radius

$$|W_i| \frac{\gamma_i + \sigma_i}{1 - \sigma_i}.$$

Remark 1.4. From the expression for the radius of inclusion disk and the fact that $\phi(\gamma_i, \sigma_i) = (\gamma_i + \sigma_i)/(1 - \sigma_i)$ is monotonically increasing function in both variables $\gamma_i, \sigma_i \in (0, 1)$, it follows that this disk is smaller if γ_i and σ_i are smaller.

Let $w = \max_{1 \leq i \leq n} |W_i|$ and let $d = \min_{i \neq j} |z_i - z_j|$ be the minimal distance between distinct approximations $z_1, \ldots, z_n$ to the zeros of P. Let us assume that the inequality

$$w < c_n d \tag{1.61}$$

is satisfied, where c_n, called i-factor (see Chap. 3), is the quantity that depends only on the polynomial degree n. The inequality (1.61) can be applied for finding the upper bound of γ_i and σ_i defined in Theorem 1.4. These bounds will depend on the i-factor c_n and our aim is to determine the maximal number c_n, so that both inequalities (1.60) are satisfied. This is the subject of the following theorem.

Theorem 1.5. *If the i-factor c_n appearing in (1.61) is not greater than $1/(2n)$, then both inequalities (1.60) hold and the minimal radius of the inclusion disk given in Theorem 1.4 is not greater than $|W_i|$.*

Proof. First, we have $|z_j - \eta_i| = |z_j - z_i + W_i| \geq |z_j - z_i| - |W_i| \geq d - w$ so that, in view of (1.61), we bound

$$\gamma_i \leq \frac{w}{d - w} < \frac{c_n}{1 - c_n}, \quad \sigma_i \leq \frac{(n-1)w}{d - w} < \frac{(n-1)c_n}{1 - c_n}.$$

Now, the inequalities (1.60) become

$$\frac{(n-1)c_n}{1 - c_n} + 2\sqrt{\frac{c_n}{1 - c_n} \cdot \frac{(n-1)c_n}{1 - c_n}} < 1, \quad \frac{c_n}{1 - c_n} + \frac{2(n-1)c_n}{1 - c_n} < 1,$$

or, after some rearrangement with the assumption $c_n < 1$,

$$c_n < \frac{1}{n + 2\sqrt{n-1}} \quad \text{and} \quad c_n < \frac{1}{2n}.$$

For $n \geq 3$, both inequalities will be satisfied if we choose $c_n < 1/(2n)$. Using the above bounds of γ_i and σ_i and Remark 1.4, we estimate

$$\frac{\gamma_i + \sigma_i}{1 - \sigma_i}|W_i| < \frac{nc_n}{1 - nc_n}|W_i| < |W_i|,$$

which proves the second part of the theorem. □

Theorem 1.6. *Let $c_n = 1/(\alpha n + \beta)$, $\alpha \geq 2$, $\beta > (2 - \alpha)n$, and let us assume that $w < c_n d$ holds. Then for $n \geq 3$, the disks*

$$D_1 := \left\{ z_1 - W_1; \frac{n}{(\alpha - 1)n + \beta}|W_1| \right\}, \ldots, D_n := \left\{ z_n - W_n; \frac{n}{(\alpha - 1)n + \beta}|W_n| \right\}$$

are mutually disjoint and each of them contains one and only one zero of P.

Proof. Using the bounds of γ_i and σ_i given in the proof of Theorem 1.5, we find

$$\gamma_i < \frac{1}{\alpha n + \beta - 1}, \quad \sigma_i < \frac{n-1}{\alpha n + \beta - 1}$$

so that for the upper bound of the inclusion radii (centered at $\eta_i = z_i - W_i$), we have

$$\frac{\gamma_i + \sigma_i}{1 - \sigma_i}|W_i| < \frac{n}{(\alpha - 1)n + \beta}|W_i|.$$

In view of Theorems 1.4 and 1.5, the inclusion disk

$$D_i = \left\{ z_i - W_i; \frac{n}{(\alpha - 1)n + \beta}|W_i| \right\} \quad (i \in \mathbf{I}_n)$$

will contain exactly one zero of P if $D_i \cap D_j = \emptyset$ ($i \neq j$). Using the inequality

$$\alpha n + \beta - 2 > 2 > 2n/((\alpha - 1)n + \beta),$$

we find

$$|\text{mid } D_i - \text{mid } D_j| = |(z_i - W_i) - (z_j - W_j)| \geq |z_i - z_j| - |W_i| - |W_j|$$
$$\geq d - 2w > w(\alpha n + \beta - 2) > \frac{n}{(\alpha - 1)n + \beta}(|W_i| + |W_j|)$$
$$= \text{rad } D_1 + \text{rad } D_2.$$

Hence, from a geometrical construction (see also (1.69)), it follows $D_i \cap D_j = \emptyset$, $i \neq j$. $\square$

Corollary 1.1. *Under the conditions of Theorem* 1.6, *each of disks* D_i^* *defined by*

$$D_i^* := \left\{ z_i; \frac{\alpha n + \beta}{(\alpha - 1)n + \beta}|W_i| \right\} = \left\{ z_i; \frac{1}{1 - nc_n}|W_i| \right\} \quad (i \in I_n)$$

contains exactly one zero of P.

Proof. Since

$$D_i = \left\{ z_i - W_i; \frac{n}{(\alpha - 1)n + \beta}|W_i| \right\} \subset \left\{ z_i; |W_i| + \frac{n}{(\alpha - 1)n + \beta}|W_i| \right\}$$
$$= \left\{ z_i; \frac{\alpha n + \beta}{(\alpha - 1)n + \beta}|W_i| \right\} = D_i^* \quad (i \in I_n)$$

and having in mind that $(\alpha - 1)n + \beta > 2$ holds under the conditions of Theorem 1.6, we have

$$|z_i - z_j| \geq d > \frac{\alpha n + \beta}{(\alpha - 1)n + \beta}(|W_i| + |W_j|) = \text{rad } D_i^* + \text{rad } D_j^* \quad (i \neq j),$$

which means that the disks $D_1^*, \ldots, D_n^*$ are also separated. Therefore, each of them contains one and only one zero of P. $\square$

The disks $D_1^*, \ldots, D_n^*$, although larger in size compared with those defined in Theorem 1.6, are centered at z_i, which is simpler for calculation. This advantage is used in Chaps. 3–5 to estimate the difference $|z_i - \zeta_i|$ ($< \text{rad } D_i^*$).

From Theorem 1.6, we see that a small c_n produces smaller inclusion disks. But in that case, the condition (1.61) becomes stronger in the sense that it requires smaller $|W_i|$. Considering the above opposite requirements for the i-factor c_n, the following natural question arises: What is more productive (1) to increase c_n in (1.61) and weaken the convergence condition or (2) to decrease c_n and produce smaller inclusion disks? To answer this question, we emphasize that weaker conditions are sufficient to ensure the convergence of the most frequently used inclusion methods, see [110]. In practice, interval methods converge starting with considerably larger disks. Besides, the need for smaller disks appears mainly to simplify theoretical analysis and avoid perplexed inequalities. Finally, it should be stressed that too small an i-factor c_n would cause serious difficulties in realization of the inequality $w < c_n d$ because small values of w assume very close initial approximations to the zeros. Altogether, the choice of as great as possible an i-factor c_n has undoubtable advantages (option (1)). This subject is also discussed in Chap. 3.

1.3 Complex Interval Arithmetic

In Chaps. 3–5, we handle complex circular intervals to obtain some estimates and bounds. For the reader's convenience, we digress briefly to list the basic properties and operations of circular complex arithmetic. For more details, see the books by Alefeld and Herzberger [3, Chap. 5] and M. Petković and L. Petković [129, Chap. 1].

A disk Z with radius $r = \operatorname{rad} Z$ and center $c = \operatorname{mid} Z \in \mathbb{C}$, where $\mathbb{C}$ is the set of complex numbers, will be denoted with $Z = \{c; r\} = \{z : |z - c| \leq r\}$. This notation provides that operations of circular arithmetic may be easily parametrized. The set of all complex circular intervals (disks) is denoted by $K(\mathbb{C})$.

We begin by defining basic circular arithmetic operations. If $0 \notin Z$ (i.e., $|c| > r$), then the inverse of Z is obtained by the Möbius transformation

$$Z^{-1} := \left\{ \frac{\bar{c}}{|c|^2 - r^2}; \frac{r}{|c|^2 - r^2} \right\} = \left\{ \frac{1}{z} : z \in Z \right\}. \tag{1.62}$$

From (1.62), we observe that the inversion Z^{-1} is an exact operation. Sometimes, we will use the so-called centered inverse of a disk,

$$Z^{I_c} := \left\{ \frac{1}{c}; \frac{r}{|c|(|c| - r)} \right\} \supseteq Z^{-1}, \tag{1.63}$$

enlarged with respect to the exact range Z^{-1}, but simpler for calculations. Actually, this is the Taylor's form of inversion derived in [105].

Furthermore, if $Z_k = \{c_k; r_k\}$ $(k = 1, 2)$, then

$$Z_1 \pm Z_2 := \{c_1 \pm c_2; r_1 + r_2\} = \{z_1 \pm z_2 : z \in Z_1, z_2 \in Z_2\}. \tag{1.64}$$

In general, the following is valid

$$\sum_{k=1}^{n} Z_k = \left\{ \sum_{k=1}^{n} c_k; \sum_{k=1}^{n} r_k \right\}.$$

If $\alpha \in \mathbb{C}$, then
$$\alpha \cdot \{c; r\} = \{\alpha c; |\alpha| r\}. \tag{1.65}$$

The product $Z_1 \cdot Z_2$ is defined as in [50]:

$$Z_1 \cdot Z_2 := \{c_1 c_2; |c_1| r_2 + |c_2| r_1 + r_1 r_2\} \supseteq \{z_1 z_2 : z_1 \in Z_1, z_2 \in Z_2\}. \tag{1.66}$$

Then, by (1.62), (1.63), and (1.66),

$$Z_1 : Z_2 := Z_1 \cdot Z_2^{-1} \quad \text{or} \quad Z_1 : Z_2 := Z_1 \cdot Z_2^{I_c} \quad (0 \notin Z_2).$$

Using the definition (1.66), we derive

$$\prod_{k=1}^{n} \{c_k; r_k\} = \left\{ \prod_{k=1}^{n} c_k; \prod_{k=1}^{n} (|c_k| + r_k) - \prod_{k=1}^{n} |c_k| \right\}. \tag{1.67}$$

For two disks $Z_1 = \{c_1; r_1\}$ and $Z_2 = \{c_2; r_2\}$, the following is valid:

$$Z_1 \subseteq Z_2 \quad \Longleftrightarrow \quad |c_1 - c_2| \leq r_2 - r_1, \tag{1.68}$$
$$Z_1 \cap Z_2 = \emptyset \quad \Longleftrightarrow \quad |c_1 - c_2| > r_1 + r_2. \tag{1.69}$$

A fundamental property of interval computation is *inclusion isotonicity*, which forms the basis for almost all applications of interval arithmetic. Theorem 1.7 shows this property for the four basic operations in circular complex arithmetic ([3, Chap. 5]).

Theorem 1.7. *Let A_k, $B_k \in K(\mathbb{C})$ $(k = 1, 2)$ be such that*

$$A_k \subseteq B_k \quad (k = 1, 2).$$

Then

$$A_1 * A_2 \subseteq B_1 * B_1$$

holds for the circular complex operation $* \in \{+, -, \cdot, :\}$.

Let f be a complex function over a given disk $Z \in K(\mathbb{C})$. The complex-valued set $\{f(z) : z \in Z\}$ is not a disk in general. To deal with disks, we introduce a *circular extension* F of f, defined on a subset $D \subseteq K(\mathbb{C})$ such that

$$F(Z) \supseteq \{f(z) : z \in Z\} \quad \text{for all } Z \in D \quad (inclusion),$$
$$F(z) = f(z) \quad \text{for all } z \in Z \quad (complex\ restriction).$$

We shall say that the complex interval extension F is *inclusion isotone* if the implication

$$Z_1 \subseteq Z_1 \quad \Longrightarrow \quad F(Z_1) \subseteq F(Z_2)$$

is satisfied for all $Z_1, Z_1 \in D$. In particular, we have

$$z \in Z \implies f(z) = F(z) \in F(Z). \tag{1.70}$$

The following simple property is valid:

$$\max\{0, |\operatorname{mid} F(Z)| - \operatorname{rad} F(Z)\} \leq |f(z)| \leq |\operatorname{mid} F(Z)| + \operatorname{rad} F(Z)$$

for all $z \in Z$. In particular, if $f(z) = z$ and $z \in Z$, then

$$\max\{0, |\operatorname{mid} Z| - \operatorname{rad} Z\} \leq |z| \leq |\operatorname{mid} Z| + \operatorname{rad} Z. \tag{1.71}$$

The square root of a disk $\{c; r\}$ in the centered form, where $c = |c|e^{i\theta}$ and $|c| > r$, is defined as the union of two disks (see Gargantini [47]):

$$\{c; r\}^{1/2} := \left\{ \sqrt{|c|}e^{i\theta/2}; \eta \right\} \bigcup \left\{ -\sqrt{|c|}e^{i\theta/2}; \eta \right\}, \tag{1.72}$$
$$\eta = \sqrt{|c|} - \sqrt{|c| - r}.$$

Chapter 2
Iterative Processes and Point Estimation Theory

This chapter is devoted to estimations of zero-finding methods from data at one point in the light of Smale's point estimation theory [165]. One of the crucial problems in solving equations of the form $f(z) = 0$ is the construction of such initial conditions, which provide the guaranteed convergence of the considered numerical algorithm. These initial conditions involve an initial approximation $z^{(0)}$ to a zero ζ of f and they should be established in such a way that the sequence $\{z^{(m)}\}_{m=1,2,\ldots}$ of approximations, generated by the implemented algorithm which starts from $z^{(0)}$, tends to the zero of f. The construction of initial conditions and the choice of initial approximations ensuring the guaranteed convergence are very difficult problems that cannot be solved in a satisfactory way in general, even in the case of simple functions, such as algebraic polynomials.

In Sect. 2.1, we present some historical data and Smale's point estimation theory applied to Newton's method. More generalized problems are discussed in Sect. 2.2, where the whole class of quadratically convergent methods is treated, and in Sect. 2.3, where Smale's work is applied to the third-order methods. Improvements of Smale's result related to Newton's method, carried out by X. Wang and Han [181] and D. Wang and Zhao [178], are the subject of Sect. 2.4. Their approach is applied to the convergence analysis of the Durand–Kerner's method for the simultaneous determination of all zeros of a polynomial (Sect. 2.5).

2.1 Newton's Method Estimates

Newton's method and its modifications have been often used for solving nonlinear equations and systems. The Newton's method attempts to solve $f(z) = 0$ by an iteratively defined sequence

$$z_{m+1} = z_m - f'(z_m)^{-1}f(z_m) \quad (m = 0, 1, \ldots)$$

M. Petković, *Point Estimation of Root Finding Methods*. Lecture Notes in Mathematics 1933,
© Springer-Verlag Berlin Heidelberg 2008

for an initial point z_0. If this initial point is well chosen, then the convergence of the sequence $\{z_m\}$ is reasonably fast.

However, not much is known about the region of convergence and it is difficult to obtain a priori knowledge of convergence. Let d denote the dimension of a system of nonlinear equations (with d unknown variables). The quadratic character of convergence in case of a single equation ($d = 1$) was discussed by J. B. J. Fourier [42]. The first convergence proof of the Newton's method for $d = 2$, which uses only the second derivatives, was given by A. M. Ostrowski in 1936. The proof for general d was presented in a doctoral thesis by Bussmann and communicated by Rechbock in the journal *Z. Angew. Math. Mech.* **22** (1942). Later, Kantorovich adapted the proof to very general problems of functional analysis [70]. From a theoretical as well as practical point of view, this fundamental result has had a great importance and it has initiated a series of papers (e.g., Kantorovich and Akilov [71], Ostrowski [99], Gragg and Tapia [53], Rall [152], Traub and Wozniakowski [173]).

Kantorovich's approach in solving systems of nonlinear equations, very influential in this area, has the following properties (a) weak differentiability hypotheses are imposed on the system, although analyticity (as a strong hypothesis) is assumed and (b) it is supposed that derivative bounds exist over the entire domain. On the other hand, Smale [165], [167] deduced necessary information from data at a single point. This original viewpoint makes the valuable advance in the theory and practice of iterative processes for solving nonlinear equations and we take Smale's point estimation theory as the main subject of this chapter.

First results concerned with computationally verifiable domain of convergence were found by Smale [165] and Shub and Smale [163] (see also Kim [73]) and they depend on classical research connected with *schlicht* function theory. Later, in 1986, Smale [167] proposed a different strategy based on "point data" to state the theorems about convergence of Newton's method in Banach space. Most of Sect. 2.2 is, essentially, the adaptation of Smale's work presented in [167].

In this chapter, we deal with the so-called *approximate zero*, defined as a point that provides fast convergence to a zero immediately after starting iterative process. Although several versions of this notion exist for Newton's method and other higher-order iterative methods (see [163]–[167]), the definition given in [167] seems most acceptable since it handles the terms of convergence property and, at the same time, points to the order of convergence. To introduce Smale's idea, we consider first the simple case of a single polynomial

$$f(z) = a_n z^n + a_{n-1} z^{n-1} + \cdots + a_1 z + a_0$$

of degree n. The Newton's method, applied to the polynomial equation $f(z) = 0$, starts with $z_0 \in \mathbb{C}$ and produces successively the sequence of approximations $\{z_m\}$ by the iterative formula

$$z_m = z_{m-1} - \frac{f(z_{m-1})}{f'(z_{m-1})} \quad (m = 0, 1, \dots).$$

Having in mind the quadratic convergence of Newton's method, Smale [167] introduced the notion of approximate zeros by the following definition.

Definition 1. $z_0 \in \mathbb{C}$ is an *approximate zero* of f if it satisfies the following convergence condition

$$\|z_m - z_{m-1}\| \leq \left(\frac{1}{2}\right)^{2^{m-1}-1} \|z_1 - z_0\| \quad (m = 1, 2, \dots).$$

Sometimes, the above inequality is written in the equivalent form

$$\left\|\frac{f(z_m)}{f'(z_m)}\right\| \leq \left(\frac{1}{2}\right)^{2^{m-1}-1} \left\|\frac{f(z_0)}{f'(z_0)}\right\| \quad (m = 1, 2, \dots).$$

We observe that the exponent in the two last inequalities indicates the quadratic convergence. Evidently, in the case of single polynomials, the norm $\|\cdot\|$ can be replaced by the absolute value $|\cdot|$.

Let $f^{(k)}$ denote the kth derivative of f at a point $z \in \mathbb{C}$. A function $\alpha(z, f)$ defined by

$$\alpha(z, f) = \left|\frac{f(z)}{f'(z)}\right| \sup_{k>1} \left|\frac{f^{(k)}(z)}{k!f'(z)}\right|^{1/(k-1)}$$

has an important role in constructing a test for an approximate zero.

Now, let $f : E \to F$ be an analytic map from an open subset of a real or complex Banach space E to a space F. A great deal of the subsequent consideration assumes the finite-dimensional cases $E = \mathbb{C}^n$, $F = \mathbb{C}^n$, where n could be equal to 1 as in the case of single polynomials. In particular, the map f could be related to a system of polynomials.

Following the notation of Kantorovich and Akilov [71] and Collatz [21], the Fréchet derivative of $f : E \to F$ at a point $z \in E$ will be denoted by $f'_{(z)}$. Furthermore, $f^{(k)}_{(z)}$ denotes the kth derivative of f at a point z in the sense of Fréchet. For brevity, we will write

$$f^{(k)}_{(z)} h^k \quad \text{instead of} \quad f^{(k)}_{(z)} \underbrace{(h, \dots, h)}_{k \text{ times}}.$$

For example, if $f : X \to X$ is an analytic mapping on $D \subset X$ and z and $z+h$ belong to D, then the Taylor's expansion of f at the point z is given by

$$f(z + h) = \sum_{k=0}^{\infty} \frac{1}{k!} f^{(k)}_{(z)} \underbrace{(h, \dots, h)}_{k \text{ times}} = \sum_{k=0}^{\infty} \frac{1}{k!} f^{(k)}_{(z)} h^k.$$

Assume that the derivative $f'_{(z)}$ is invertible, then Newton's method generates a new vector $\hat{z}$ from z by

$$\hat{z} = z - \left(f'_{(z)}\right)^{-1} f(z).$$

Let us denote the norm of this Newton's step $\hat{z} - z$ with β, i.e.,

$$\beta = \beta(z, f) = \|\left(f'_{(z)}\right)^{-1} f(z)\|.$$

If $f'_{(z)}$ is not invertible, then we let $\beta = \infty$.

Starting from a point $z_0 \in E$, Newton's method generates the sequence of successive approximations $z_m = z_{m-1} - \left(f'_{(z_{m-1})}\right)^{-1} f(z_{m-1})$ (if possible) in Banach space. Following Definition 1, the point $z_0 \in E$ is called an *approximate zero* of f if z_m is defined for all m and satisfies:

$$\|z_m - z_{m-1}\| \le \left(\frac{1}{2}\right)^{2^{m-1}-1} \|z_1 - z_0\|, \quad \text{for all} \ \ m \ge 1.$$

Hence, it follows that $\{z_m\}$ is a Cauchy sequence with a limit, say $\zeta \in E$. It is easy to see that $f(\zeta) = 0$. Indeed, since $z_{m+1} = z_m - \left(f'_{(z_m)}\right)^{-1} f(z_m)$, we estimate

$$\|f(z_m)\| = \|f'_{(z_m)}(z_{m+1} - z_m)\| \le \|f'_{(z_m)}\| \, \|z_{m+1} - z_m\|.$$

Taking the limit as $m \to \infty$ yields

$$\|f(\zeta)\| \le \|f'_{(z_m)}\| \lim_{m \to \infty} \|z_{m+1} - z_m\| = 0.$$

Having in mind Definition 1, we note that for an approximate zero, Newton's method is superconvergent beginning with the first iteration.

To give criteria for z to be an approximate zero, let us define

$$\gamma(z, f) = \sup_{k>1} \left\| \left(f'_{(z)}\right)^{-1} \frac{f^{(k)}_{(z)}}{k!} \right\|^{1/(k-1)}$$

or, if $\left(f'_{(z)}\right)^{-1}$ or the *supremum* does not exist, set $\gamma(z, f) = \infty$. The norm is assumed to be a norm of a multilinear map as it is defined in [29] and [83]. It will be convenient sometimes to write $\gamma(z, f)$ as $\gamma_f(z)$, $\gamma(z)$, or γ. Finally, let us define $\alpha(z, f) = \beta(z, f)\gamma(z, f)$, where β was defined earlier.

Assume now that $f : E \to F$ is a map expressed as $f(z) = \sum_{k=0}^{n} a_k z^k$, where $z \in E$, $0 < n \le \infty$. Here, E and F are Banach spaces, and a_k is a bounded symmetric k-linear map from $E \times \cdots \times E$ (k times) to F. In other words, $a_k z^k$ is a homogeneous polynomial of degree k. For $E = \mathbb{C}^d$, we meet the case in the usual sense, and for $d = 1$ (one variable) a_k is kth coefficient (real or complex) of f. If n is finite, then f is a polynomial.

Let us define
$$\|f\| = \sup_{k \geq 0} \|a_k\| \quad \text{(general case)},$$

where $\|a_k\|$ is the norm of a_k as a bounded map, and introduce

$$\phi_n(r) = \sum_{i=0}^{n} r^i, \quad \phi_n'(r) = \frac{d}{dr}\phi_n(r), \quad \phi(r) = \phi_\infty(r) = \frac{1}{1-r}.$$

The main result will be stated through some lemmas. Let us suppose that E and F are Banach spaces, both real or both complex.

Lemma 2.1. *Let $A, B : E \to F$ be bounded linear maps and let A be invertible. Assume that the inequality $\|A^{-1}B - I\| < c < 1$ holds, then B is invertible and $\|B^{-1}A\| < 1/(1-c)$.*

Proof. Let $V = I - A^{-1}B$. Since $\|V\| < 1$, $\displaystyle\sum_{k=0}^{\infty} V^k$ exists and its norm is less than $1/(1-c)$. Furthermore,

$$(I - V) \sum_{k=0}^{n} V^k(I - V) = I - V^{n+1}.$$

By taking limits, we observe that $A^{-1}B = I - V$ is invertible with the inverse $\displaystyle\sum_{k=0}^{\infty} V^k$. Note that $A^{-1}B \displaystyle\sum_{k=0}^{\infty} V^k = I$, and hence B can be expressed as the composition of invertible maps. $\square$

Lemma 2.2. *Assume that $f : E \to F$ is analytic and $\hat{z}, z \in E$ are such that $\|\hat{z} - z\| < 1 - \sqrt{2}/2$. Then:*

(i) $f'_{(\hat{z})}$ *is invertible.*

(ii) $\left\| \left(f'_{(\hat{z})}\right)^{-1} f'_{(z)} \right\| < \dfrac{1}{2 - \phi'(\|\hat{z} - z\|\gamma(z))}.$

(iii) $\gamma(\hat{z}) \leq \gamma(z)\dfrac{1}{2 - \phi'(\|\hat{z} - z\|\gamma(z))} \left(\dfrac{1}{1 - \|\hat{z} - z\|\gamma(z)} \right)^3.$

Proof. Using the Taylor's expansion of f' in the neighborhood of the point z, one obtains

$$f'_{(\hat{z})} = \sum_{k=0}^{\infty} \frac{f_{(z)}^{(k+1)}}{k!}(\hat{z} - z)^k.$$

Hence,

$$\left(f'_{(z)}\right)^{-1} f'_{(\hat{z})} = \sum_{k=0}^{\infty} \left(f'_{(z)}\right)^{-1} \frac{f_{(z)}^{(k+1)}}{k!}(\hat{z} - z)^k$$

and

$$\left(f'_{(z)}\right)^{-1} f'_{(\hat{z})} - I = \sum_{k=1}^{\infty} (k+1) \frac{\left(f'_{(z)}\right)^{-1} f_{(z)}^{(k+1)}}{(k+1)!} (\hat{z} - z)^k.$$

According to this, we have

$$\left\| \left(f'_{(z)}\right)^{-1} f'_{(\hat{z})} - I \right\| \leq \sum_{k=1}^{\infty} (k+1) \left\| \frac{\left(f'_{(z)}\right)^{-1} f_{(z)}^{(k+1)}}{(k+1)!} \right\| \|\hat{z} - z\|^k$$

$$\leq \sum_{k=1}^{\infty} (k+1) \left(\gamma(z) \|\hat{z} - z\| \right)^k$$

$$\leq \phi'(\gamma(z) \|\hat{z} - z\|) - 1.$$

Since $\gamma(z) \|\hat{z} - z\| < 1 - \sqrt{2}/2$, all the series in this proof are convergent. Besides, for $r < 1 - \sqrt{2}/2$, we have $\phi'(r) - 1 < 1$, where $\phi'(r) = 1/(1 - r)^2$. According to Lemma 2.1, the assertions (i) and (ii) of Lemma 2.2 follow.

Starting from $\phi(r) = 1/(1 - r)$, we derive

$$\frac{\phi^{(\nu)}(r)}{\nu!} = \sum_{k=0}^{\infty} \binom{\nu + k}{k} r^k = \frac{1}{(1 - r)^{\nu+1}}.$$

Also, we have

$$\frac{1}{2 - \phi'(r)} \cdot \frac{1}{(1 - r)^2} = \frac{1}{\psi(r)},$$

where

$$\psi(r) = 2r^2 - 4r + 1.$$

The above formulae, including the derivatives of ϕ, will be used in further consideration.

To prove (iii) of Lemma 2.2, we put

$$\gamma_k = \gamma_k(z) = \left\| \left(f'_{(z)}\right)^{-1} \frac{f_{(z)}^{(k)}}{k!} \right\|^{1/(k-1)} \qquad \text{and} \qquad \gamma = \sup_{k>1} \gamma_k.$$

Then, by Taylor's expansion, we estimate

$$\gamma_k(\hat{z})^{k-1} = \left\| \left(f'_{(z)}\right)^{-1} f'_{(z)} \sum_{\nu=0}^{\infty} \frac{\left(f'_{(z)}\right)^{-1} f_{(z)}^{(k+\nu)} (\hat{z} - z)^{\nu}}{\nu! k!} \right\|$$

$$\leq \left\| \left(f'_{(z)}\right)^{-1} f'_{(z)} \right\| \sum_{\nu=0}^{\infty} \binom{k+\nu}{\nu} \left\| \frac{\left(f'_{(z)}\right)^{-1} f_{(z)}^{(k+\nu)} (\hat{z} - z)^{\nu}}{(k+\nu)!} \right\|$$

$$\leq \left\|\left(f'_{(z)}\right)^{-1}f'_{(z)}\right\|\gamma(z)^{k-1}\sum_{\nu=0}^{\infty}\binom{k+\nu}{\nu}(\|\hat{z}-z\|\gamma(z))^{\nu}$$

$$\leq \left\|\left(f'_{(z)}\right)^{-1}f'_{(z)}\right\|\gamma^{k-1}\left(\frac{1}{1-\gamma\|\hat{z}-z\|}\right)^{k+1}.$$

Using (ii) of Lemma 2.2 and extracting the $(k-1)$th root, we obtain

$$\gamma_k(\hat{z}) \leq \frac{\gamma}{2-\phi'(\gamma\|\hat{z}-z\|)}\left(\frac{1}{1-\gamma\|\hat{z}-z\|}\right)^{(k+1)/(k-1)}.$$

The supremum is achieved at $k=2$, which leads to the statement (iii) of Lemma 2.2. $\square$

Lemma 2.3. (i) *Let* $\alpha = \alpha(z,f) < 1$ *and* $\hat{z} = z - \left(f'_{(z)}\right)^{-1}f(z)$, $\beta = \beta(z,f)$. *Then, we have*

$$\left\|\left(f'_{(z)}\right)^{-1}f(\hat{z})\right\| \leq \beta\left(\frac{\alpha}{1-\alpha}\right).$$

(ii) *Let* $z, \hat{z} \in E$ *with* $f(z) = 0$, *and let the inequality* $\|\hat{z}-z\|\gamma(z) < 1$ *hold. Then,*

$$\left\|\left(f'_{(z)}\right)^{-1}f(\hat{z})\right\| \leq \frac{\|\hat{z}-z\|}{1-\|\hat{z}-z\|\gamma(z)}.$$

Proof. Using the Taylor's series, we obtain

$$\left(f'_{(z)}\right)^{-1}f(\hat{z}) = \sum_{k=0}^{\infty}\left(f'_{(z)}\right)^{-1}\frac{f^{(k)}_{(z)}}{k!}(\hat{z}-z)^k.$$

We omit the first two terms on the right side because of

$$\left(f'_{(z)}\right)^{-1}f(z) + \left(f'_{(z)}\right)^{-1}f'_{(z)}(\hat{z}-z) = \underbrace{(z-\hat{z})}_{Newton's\ step} + (\hat{z}-z) = 0.$$

Since $\beta(z) = \|\hat{z}-z\|$, there follows

$$\left\|\left(f'_{(z)}\right)^{-1}f(\hat{z})\right\| \leq \beta\sum_{k=2}^{\infty}(\gamma\beta)^{k-1} \leq \beta\frac{\alpha}{1-\alpha},$$

ending the proof of (i). To prove (ii), we begin as above and drop out the first term since $f(z) = 0$. Thus, we estimate

$$\left\|\left(f'_{(z)}\right)^{-1}f(\hat{z})\right\| \le \|\hat{z}-z\|\left(1+\sum_{k=2}^{\infty}\gamma_k^{k-1}\|\hat{z}-z\|^{k-1}\right)$$

$$\le \|\hat{z}-z\|\sum_{k=0}^{\infty}(\gamma\|\hat{z}-z\|)^k$$

$$\le \|\hat{z}-z\|\frac{1}{1-\|\hat{z}-z\|\gamma(z)},$$

which completes the proof of Lemma 2.3. □

Lemma 2.4. *Let* $f : E \to F$ *be an analytic map from the Banach space* E *to* F *and let* $\alpha = \alpha(z, f)$, $\beta = \beta(z)$, *and* $\hat{\beta} = \beta(\hat{z})$:

(i) *If* $\alpha < 1 - \sqrt{2}/2$, *then*

$$\hat{\beta} \le \beta\left(\frac{\alpha}{1-\alpha}\right)\left(\frac{1}{2-\phi'(\alpha)}\right).$$

(ii) *If* $f(z) = 0$ *and* $\gamma\|\hat{z}-z\| < 1 - \sqrt{2}/2$, *then*

$$\hat{\beta} \le \|\hat{z}-z\|\frac{1}{2-\phi'(\gamma\|\hat{z}-z\|)}\left(\frac{1}{1-\gamma\|\hat{z}-z\|}\right).$$

Proof. We estimate

$$\beta(\hat{z}) = \left\|\left(f'_{(\hat{z})}\right)^{-1}f(\hat{z})\right\| = \left\|\left(f'_{(\hat{z})}\right)^{-1}f'_{(z)}\left(f'_{(z)}\right)^{-1}f(\hat{z})\right\|$$

$$\le \left\|\left(f'_{(\hat{z})}\right)^{-1}f'_{(z)}\right\|\left\|\left(f'_{(z)}\right)^{-1}f(\hat{z})\right\|$$

$$\le \left(\frac{1}{2-\phi'(\alpha)}\right)\beta\left(\frac{\alpha}{1-\alpha}\right).$$

In the last step, we use the assertions (ii) of Lemma 2.2 and (i) of Lemma 2.3. The assertion (ii) of the lemma is proved in a similar way. If $f(z) = 0$, then by Lemmas 2.2(ii) and 2.3(ii), we obtain

$$\beta(\hat{z}) \le \left\|\left(f'_{(\hat{z})}\right)^{-1}f'_{(z)}\right\|\left\|\left(f'_{(z)}\right)^{-1}f(\hat{z})\right\|$$

$$\le \frac{1}{2-\phi'(\|\hat{z}-z\|\gamma(z))}\|\hat{z}-z\|\frac{1}{(1-\|\hat{z}-z\|\gamma(z))}. □$$

Lemma 2.5. *Let* $\psi(r) = 2r^2 - 4r + 1$ *as above. Using the notation of Lemma 2.4, we have the following assertions:*

(i) *If* $\alpha < 1 - \sqrt{2}/2$, *then*

$$\hat{\alpha} = \alpha(\hat{z}, f) \le \left(\frac{\alpha}{\psi(\alpha)}\right)^2.$$

(ii) *If $f(\zeta) = 0$ and $\|z - \zeta\|\gamma(\zeta) < 1 - \sqrt{2}/2$, then*

$$\hat{\alpha} \leq \frac{\gamma(\zeta)\|z - \zeta\|}{\psi(\gamma(\zeta)\|z - \zeta\|)^2}.$$

Proof. Since $\hat{\alpha} = \hat{\beta}\hat{\gamma}$, using Lemmas 2.2(iii) and 2.4(i), we get

$$\hat{\alpha} \leq \beta \left(\frac{\alpha}{1 - \alpha}\right)\left(\frac{1}{2 - \phi'(\alpha)}\right)\gamma\left(\frac{1}{1 - \alpha}\right)^3\left(\frac{1}{2 - \phi'(\alpha)}\right),$$

and hence

$$\hat{\alpha} < (\beta\gamma)\alpha\left(\frac{1}{\psi(\alpha)}\right)^2 \leq \left(\frac{\alpha}{\psi(\alpha)}\right)^2.$$

The proof of Lemma 2.5(ii) goes in a similar way using Lemmas 2.2(iii) and 2.4(ii). □

The following assertion leads to the main result.

Proposition 2.1. *Let $f : E \to F$ be analytic, $z = z_0 \in E$, and $\alpha(z)/\psi(\alpha(z))^2 = a < 1$. Let $\alpha_m = \alpha(z_m)$, $\psi_m = \psi(\alpha(z_m))$ $(m = 1, 2, \ldots)$, where $z_m = z_{m-1} - \left(f'_{(z_{m-1})}\right)^{-1}f(z_{m-1})$ $(m = 1, 2, \ldots)$, then:*

(i) $\alpha_m \leq a^{2^m-1}\alpha(z_0)$ $(m = 1, 2, \ldots)$.
(ii) z_m *is defined for all m.*
(iii) $\|z_m - z_{m-1}\| < a^{2^{m-1}-1}\|z_1 - z_0\|$, *for all m.*

Proof. Assume that a constant $A > 0$ and $a_i > 0$ satisfy the inequality $a_{i+1} \leq Aa_i^2$ for all $i = 0, 1, \ldots$. We prove by induction that

$$a_m \leq \left(Aa_0\right)^{2^m-1}a_0 \quad (m = 1, 2, \ldots).$$

According to this and Lemma 2.5(i), we proved (i), while the assertion (ii) follows from (i). Therefore, it remains to prove (iii). The case $m = 1$ is evident so that we consider the case $m > 1$.

With regard to Proposition 2.1(i) and the relation between ϕ' and ψ (see the proof of Lemma 2.2), we write

$$\|z_m - z_{m-1}\| \leq \|z_{m-1} - z_{m-2}\|\frac{\alpha_{m-2}(1 - \alpha_{m-2})}{\psi_{m-2}}.$$

Using the assertion (ii) and induction on this inequality, we get

$$\|z_m - z_{m-1}\| \leq a^{2^{m-2}-1}\|z_1 - z_0\|a^{2^{m-1}-1}\alpha(z_0)\left(\frac{1 - \alpha_{m-2}}{\psi_{m-2}}\right)$$

$$\leq a^{2^{m-1}-1}\|z_1 - z_0\|\frac{\alpha(z_0)}{\psi_{m-2}} < a^{2^{m-1}-1}\|z_1 - z_0\|,$$

where the inequality

$$\frac{\alpha(z_0)}{\psi_{m-2}} \leq \frac{\alpha_0}{\psi_0} \leq a\psi_0 < 1$$

is used. □

Now the main result, due to Smale [167], may be stated.

Theorem 2.1. *There is a constant α_0 approximately equal to 0.130707 such that if $\alpha(z, f) < \alpha_0$, then z is an approximate zero of f.*

Proof. Consider the previously introduced polynomial $\psi(r) = 2r^2 - 4r + 1$, and the function $(\alpha/\psi(\alpha))^2$ of Lemma 2.5(i). We restrict our attention to the interval determined by $0 \leq r \leq 1 - \sqrt{2}/2 \cong 0.293$. The function $r/\psi(r)^2$ increases from 0 to ∞ as r goes from 0 to $1 - \sqrt{2}/2$. In fact, Theorem 2.1 is a consequence of Proposition 2.1 where $a = 1/2$. For this reason, let α_0 be the unique r such that $r/\psi(r)^2 = a = 1/2$. Thus, α_0 is a zero of the real quartic polynomial $\psi(r)^2 - 2r = 4r^4 - 16r^3 + 20r^2 - 10r + 1$. Using Newton's method with the starting point in the interval $(0, 0.293)$, we calculate approximately $\alpha_0 = 0.130707$ and complete the proof. □

Remark 2.1. Kim [73] has independently given a proof of Theorem 2.1 with $\alpha_0 = 1/54$. Since her proof uses the theory of schlicht functions, it cannot be extended to several variables. The same author [74] has improved this bound to $\alpha_0 = 1/48$, but only for single polynomials.

Remark 2.2. Theorem 2.1 can be slightly sharpened in cases where f is a polynomial map $E \to F$ of Banach spaces of degree $n < \infty$. To confirm that, it is necessary to replace $\phi(r)$ by $\phi_n(r) = 1 + r + \cdots + r^n$ everywhere in the proofs and conclusions.

Remark 2.3. The following shows that α_0 must be less than or equal to $3 - 2\sqrt{2} \approx 0.1716$ in Theorem 2.1. Let $f_a : \mathbb{C} \to \mathbb{C}$ be $f_a(z) = 2z - z/(1 - z) - a, a > 0$. Then, $\alpha(0, f_a) = a$ and $f_a(\zeta) = 0$, where $\zeta = \left((1 + a) \pm \sqrt{(1 + a)^2 - 8a}\right)/4$. If $\alpha = a > 3 - 2\sqrt{2}$, these roots are not real, so that Newton's method for solving $f_a(z) = 0$, starting at $z_0 = 0$, will never converge.

2.2 Approximate Zeros of Quadratically Convergent Methods

Continuing in the spirit of the works by Smale [167], Curry [25], and Kim [74], Chen [19] has dealt with analytic functions in the complex plane $\mathbb{C}$ and the general Newton-like quadratically convergent iterative methods of the form $z + M(f(z), f'(z))$, which evidently include Newton's method $z - f(z)/f'(z)$ as a special case. The conditions for a point to be an approximate zero presented in [25], [74], [167] are essentially the same. They are based on the inequality

$$\frac{\gamma_f(z_0)|f(z_0)|}{|f'(z_0)|} < \alpha_0$$

for a certain constant α_0, which takes different values depending on the iterative algorithm under investigation (Newton's, Euler's, or Halley's). Here, $\gamma_f(z) = \gamma(z, f)$ is defined as in Sect. 2.1. In [19], Chen has established a similar condition for a point to be an approximate zero and shown what characteristic of the methods affects the bound. This generalization represents the original conditions for Newton's method found by Smale [167] (see also [25], [74]) in a more general frame. Furthermore, Chen's approach enables that other iterative methods of this form can be applied to improve the convergence behavior of Newton's method in a global sense.

Applying the iteration of a rational map $z + M(f(z), f'(z))$, each attracting cycle or fixed point attracts at least one critical point (following the Fatou–Julia's theory; see, e.g., [36], [104]). The divergent behavior appears due to the existence of attracting cycles other than the fixed points. In some situations, it is possible to perturb Newton's method in such suitable ways that some attracting cycles are removed. In this manner, in the region where Newton's method diverges, one may alter it to some other methods to accomplish convergence. Chen [19] presented the following example: Newton's method when applied to the polynomial $f(z) = \frac{1}{2}z^3 - z + 1$ has the attracting cycle $\{0; 1\}$ on the real line. When using

$$M(f, f') = -\frac{f}{f'} + \frac{f^3}{f'(1 + 2f^2)} \quad (f \equiv f(z)),$$

the attracting cycle $\{0; 1\}$ is removed. Moreover, the whole real line is free of any attracting cycles.

In this section, we give a short review (without proofs) of Chen's results [19] concerning approximate zeros of quadratically convergent methods. Let f be any analytic function in the complex plane $\mathbb{C}$ and define a function $M : \mathbb{C} \times (\mathbb{C} \setminus \{0\}) \to \mathbb{C}$. In what follows, we will study iterative methods for finding zeros of f in the form

$$I_f(z) = z + M(f(z), f'(z)).$$

As usual, we write I_f^n for the n-fold composition $I_f \circ \cdots \circ I_f$ of I_f, so that $I_f^n(z)$ is the nth iterate $I_f(I_f(\cdots(I_f(z))))$ of z.

Assume that M satisfies the following:

(a) $M : \mathbb{C} \times (\mathbb{C} \setminus \{0\}) \to \mathbb{C}$ is analytic;

(b) $M(0, \cdot) \equiv 0$;

(c) $M^{(1,0)}(0, v) = -\frac{1}{v}$, where $M^{(1,0)}(u, v) = \frac{\partial M(u, v)}{\partial u}$.

Assumptions (b) and (c) mean that the iteration I_f is locally quadratically convergent at simple zeros of f. In particular, Newton's method is $M(u,v) = -u/v$, which evidently satisfies assumptions (a)–(c).

Instead of Smale's Definition 1, Chen [19] defines the notion of an approximate zero in terms of the residual and in a somewhat stronger sense.

Definition 2. Let I_f be the iterative method defined on M and let $z_m = I_f^m(z_0)$ for $m \geq 0$. We call z_0 an *approximate zero* of f with respect to I_f if there exist positive numbers a_0, a_1, $a_2, \ldots$ satisfying $a_0 < 1$ and $a_{k+1} \leq a_k^2$ for all $k \geq 0$ such that $|f(z_{m+k})| < a_k^{2^m-1}|f(z_k)| \quad (m \geq 0, \ k \geq 0)$.

Remark 2.4. If z_0 is an approximate zero, then so is $z_1 = I_f(z_0)$, which means that Definition 2 of an approximate zero is forward invariant. Note that Smale's Definition 1 does not have this property. Another advantage is that it directly points to the quadratic convergence since, assuming that z_0 is an approximate zero,

$$|f(z_m)| \leq a_0^{2^m-1}|f(z_0)| \quad (m \geq 0),$$

which is the residual version of the original definition in [167] taking $a_0 = 1/2$. It is clear that the condition for a point to be an approximate zero found in [167] is still valid under Definition 2.

Let

$$\gamma_f(z) = \sup_{k>1} \left| \frac{f^{(k)}(z)}{k! f'(z)} \right|^{1/(k-1)},$$

$$\tau_M(u,v) = \sup_{n_1+n_2 \geq 1} \left| \frac{M^{(n_1,n_2)}(u,v)}{n_1! n_2!} \right|^{1/(n_1+n_2)},$$

where $M^{(n_1,n_2)}(u,v) = \partial^{n_1+n_2} M(u,v)/\partial u^{n_1} \partial v^{n_2}$. Chen [19] has stated the following theorem.

Theorem 2.2. *Assume that M satisfies the assumptions (a)–(c) and let*

$$K_M \stackrel{\text{def}}{=} \sup_{|v| \neq 0} |v| \tau_M(0,v) < \infty.$$

Then, there exists a number $t_1(K_M)$ depending on K_M such that if

$$(\gamma_f(z_0) + 1)|f(z_0)| \tau_M(0, f'(z_0)) < t_1(K_M),$$

then z_0 is an approximate zero of f with respect to M, i.e., there exist positive numbers $a_0, a_1, a_2, \ldots$ satisfying $a_0 < 1$ and $a_{k+1} \leq a_k^2$ for all $k \geq 0$ such that

$$|f(z_{m+k})| \leq a_k^{2^m-1}|f(z_k)|$$

for all $m \geq 0$ and $k \geq 0$, where $z_m = I_f^m(z_0)$.

Let $t_1^*(K_M) = t_1(K_M)/K_M$. Then, we immediately obtain corollary to Theorem 2.2.

Corollary 2.1. *If the inequality*

$$(\gamma_f(z_0) + 1)\left|\frac{f(z_0)}{f'(z_0)}\right| < t_1^*(K_M)$$

is valid, then z_0 is an approximate zero of f with respect to M.

As a special case, consider Newton's method. One computes $\tau_M(0, v) = 1/|v|$ so that $K_M = 1$ and, consequently, $t_1^*(1) = t_1(1) \approx 0.142301$. Therefore, we have the following result.

Corollary 2.2. *If*

$$(\gamma_f(z_0) + 1)\left|\frac{f(z_0)}{f'(z_0)}\right| < t_1(1) \approx 0.142301,$$

then z_0 is an approximate zero of f with respect to the Newton's iteration.

It has been shown in [19] that the functions $t_1(K_M)$ and $t_1^*(K_M)$ are decreasing in K_M. According to this, we can conclude that the bound in the condition for approximate zeros depends on the algorithm M through the number K_M in a decreasing manner.

Let us introduce

$$T_f(z) = (\gamma_f(z) + 1)|f(z)|\tau_M(0, f'(z))$$

and

$$\Psi(x, K_M) = \frac{K_M x(1 - x)^2}{[(1 + K_M)(1 - x)^2 - K_M]^2[2(1 - x)^2 - 1]}.$$

Chen [19, Lemma 3.10] has proved that $a_0 = \Psi(T_f(z_0), K_M)$. Therefore, to have $a_0 \le 1/2$ as in Smale's definition of an approximate zero (Definition 1 in Sect. 2.1) for Newton's method, it is necessary that the inequality

$$(\gamma_f(z_0) + 1)|f(z_0)/f'(z_0)| \le \alpha \approx 0.115354$$

holds, where α satisfies $\Psi(\alpha, 1) = 1/2$. We note that the constant bound $\alpha \approx 0.115354$ obtained from Chen's more general result is smaller than $\alpha_0 \approx 0.130707$ appearing in Theorem 2.1. This is a consequence of Chen's more general technique and the fact that Definitions 1 and 2 of an approximate zero, upon which the results were derived, are not quite the same.

As mentioned in [19], experimental work shows that the theoretical bounds given in the theorems are rather conservative, which was confirmed by Curry and Van Vleck [26] about Smale's bound for Newton's method when applied to cubic polynomials. On the other hand, these bounds are involved in sufficient conditions that provide (in a theoretical sense) a guaranteed convergence of any given quadratically convergent iteration scheme M applied to any given analytic function.

Numerical experiments performed by Chen [19] show that the regions of approximate zeros with respect to the iteration defined by $M(u,v) = -u/v + u^2/v^2$ are completely contained in the corresponding regions with respect to Newton's iteration. This fact suggests the following question: Is the Newton's iteration optimal compared with other quadratically convergent iterations in the sense of having the largest regions of approximate zeros?

2.3 Approximate Zeros of Higher-Order Methods

Smale's work [167] presented in Sect. 2.1 has been generalized by Curry [25] to higher-order methods. Curry has considered a family of iterative methods which were referred to as "incremental Euler's methods" by Shub and Smale [163]. If S^2 is the Riemann sphere, an incremental algorithm is a mapping

$$I_{h,f} : S^2 \to S^2, \quad \hat{z} = I_{h,f}(z)$$

parametrized by the variable h, $0 \leq h \leq 1$, depending on a complex polynomial f.

In this section, we present only the main results (without derivations and proofs) stated by Curry in [25]. We use the same notation established in Shub and Smale [163] and Smale [167] and used in the previous sections. Let $f(z)$ be a polynomial of degree n with $z \in \mathbb{C}$ such that both $f(z)$ and $f'(z)$ are nonzero. Denote the inverse to f by f_z^{-1} and the radius of convergence of f_z^{-1} by $r(z,f) = |f(z) - f(\theta_*)|$ for some θ_*, such that $f'(\theta_*)$ vanishes, in other words, θ_* is a critical point of f.

If $f(z) \neq 0$, let us define

$$h_1(z,f) = \frac{r(z,f)}{|f(z)|},$$

i.e.,

$$h_1(z,f) \geq \min_{\substack{\theta \\ f'(\theta)=0}} \left| \frac{f(z) - f(\theta)}{f(z)} \right|.$$

In what follows, the quantity h_1 plays an important role.

Assume $h < h_1(z,f)$, then $\hat{z} = f_z^{-1}((1-h)f(z))$ is a solution to the equation

$$\frac{f(\hat{z})}{f(z)} = 1 - h.$$

The Taylor's expansion of $f(\hat{z})$ at the point z yields

$$E_\infty(z, h, f) = \hat{z} = z + \sum_{\nu=1}^{\infty} \frac{(f_z^{-1})^{(\nu)}(f(z))(-hf(z))^\nu}{\nu!}.$$

The kth incremental Euler's algorithm $E_k(z, h, f)$ or briefly E_k is defined by

$$E_k(z, h, f) = \tau_k f_z^{-1}(1 - hf(z)) = z + \sum_{\nu=1}^{k} \frac{(f_z^{-1})^{(\nu)}(f(z))(-hf(z))^\nu}{\nu!},$$

where τ_k is the truncation operator. The order of convergence of the kth incremental Euler's methods is $k + 1$.

Let us introduce

$$\sigma_k = (-1)^{k-1} \frac{f^{(k)}(z)}{k! f'(z)} \left(\frac{f(z)}{f'(z)} \right)^{k-1}.$$

In particular, for $h = 1$, we have a basic sequence $E_k(z, 1, f)$ that was considered by Schröder [160] and, much later, by Traub [172]. For example, we obtain

$$E_1(z, 1, f) = z - \frac{f(z)}{f'(z)} \quad \text{(Newton's method)},$$

$$E_2(z, 1, f) = z - \frac{f(z)}{f'(z)}(1 - \sigma_2)$$

$$= z - \frac{f(z)}{f'(z)} \left(1 + \frac{f(z)f''(z)}{2f'(z)^2} \right) \quad \text{(Euler–Chebyshev's method)},$$

etc., and for $k = 4$ and an arbitrary h

$$E_4(z, h, f) = z - \frac{f(z)}{f'(z)} \left(h - \sigma_2 h^2 + (2\sigma_2^2 - \sigma_3)h^3 - (5\sigma_2^3 - 5\sigma_2\sigma_3 + \sigma_4)h^4 \right).$$

The following definition used in [25] and this section is a generalization of Smale's Definition 1 given in Sect. 2.1.

Definition 3. $z_0 \in \mathbb{C}$ is an *approximate zero* of a polynomial f for a zero-finding method of order k if it satisfies the following convergence condition

$$\left| \frac{f(z_m)}{f'(z_m)} \right| \le \left(\frac{1}{2} \right)^{k^m - 1} \left| \frac{f(z_0)}{f'(z_0)} \right| \quad (m = 1, 2, \ldots),$$

where $z_m = \phi_k(z_{m-1})$ and ϕ_k is a method of order k.

Let $\alpha = \alpha(z, f)$ be defined as in Sect. 2.1. Curry [25] has proved the following assertion.

Proposition 2.2. *Let f be a polynomial of degree n with $f(z), f'(z)$ nonzero, where $z \in \mathbb{C}$, and let $\hat{z} = E_k(z, h, f)$. If $\alpha(z, f) < 1$, then a constant C_k exists such that*

$$\alpha(\hat{z}, f) \leq \frac{\alpha^{k+1} C_k}{\psi^2(\alpha x)},$$

where $x = (h_1/4)(e^{4/h_1} - 1)$ and $\psi(\alpha) = 2\alpha^2 - 4\alpha + 1$.

For example, we have $C_1 = 1$, $C_2 = 2$, $C_3 = 5$, $C_4 = 16$, and $C_5 = 61$. Note that Proposition 2.2 asserts that the kth incremental Euler's algorithm is of order $k + 1$ provided α is "sufficiently small."

Let $\alpha_m = \alpha(z_m, f)$ and $\psi_m = \psi(\alpha_m x)$, where $z_m = E_k(z_{m-1}, 1, f)$. The following convergence theorem concerned with approximate zero and initial conditions for methods of order $k + 1$ has been stated in [25].

Theorem 2.3. *Let $\hat{z} = E_k(z, 1, f)$ be the kth Euler's algorithm of order $k+1$. There is an α_0 such that if $\alpha(z, f) < \alpha_0$, then the iteration scheme starting from $z = z_0$ converges and*

$$\left| \frac{f(z_m)}{f'(z_m)} \right| \leq \left(\frac{1}{2} \right)^{(k+1)^m - 1} \left| \frac{f(z_0)}{f'(z_0)} \right|$$

holds, where $(\alpha_0/\psi_0) C_k^{1/k} \leq \frac{1}{2}$.

We end this section with convergence theorems for two methods of practical interest, both cubically convergent. The proof of these theorems can be found in [25]. The methods are the second incremental Euler's method

$$E_2(z, 1, f) = \hat{z} = z - \frac{f(z)}{f'(z)} \left(1 + \frac{f(z)f''(z)}{f'(z)^2} \right), \tag{2.1}$$

often called the Euler–Chebyshev's method and the Halley's method given by

$$\hat{z} = z - \frac{f(z)}{f'(z)} \left(\frac{1}{1 - f'(z)f''(z)/2(f'(z))^2} \right) \tag{2.2}$$

(see special cases given above).

Let $D(\alpha) = (1 - 2\alpha)(1 - 6\alpha + 7\alpha^2)$.

Theorem 2.4 (Halley's method). *There is an $\alpha_0(z, f)$ such that if $\alpha(z, f) < \alpha_0$, then z_0 is an approximate zero of Halley's method (2.2) and*

$$\left| \frac{f(z_m)}{f'(z_m)} \right| \leq \left(\frac{1}{2} \right)^{3^m - 1} \left| \frac{f(z_0)}{f'(z_0)} \right|,$$

for all m, where α_0 satisfies to the inequality $\alpha_0/D(\alpha_0) \leq 1/2\sqrt{2}$ ($\alpha_0 < 0.11283$).

Let $\alpha_m = \alpha(z_m, f)$ and $\widehat{D}_m = \widehat{D}(\alpha_m) = 2\alpha_m^2(1 + \alpha_m)^2 - 4\alpha_m(1 + \alpha_m) + 1$.

Theorem 2.5 (Euler–Chebyshev's method $E_2(z, 1, h)$). *There is an* $\alpha_0(z, f)$ *such that if* $\alpha(z, f) < \alpha_0(z, f)$, *then the method* (2.1) *starting from* $z = z_0$ *converges and*

$$\left| \frac{f(z_m)}{f'(z_m)} \right| \leq \left(\frac{1}{2} \right)^{3^m - 1} \left| \frac{f(z_0)}{f'(z_0)} \right|,$$

where α_0 *satisfies the inequality* $\alpha_0 / \widehat{D}_0 \leq 1/2\sqrt{5}$ ($\alpha_0 < 0.11565$).

2.4 Improvement of Smale's Result

In Sects. 2.1 and 2.2, we have considered the convergence of Newton's method

$$z^{(m+1)} = z^{(m)} - \left(f'_{(z^{(m)})} \right)^{-1} f(z^{(m)}) \quad (m = 0, 1, 2 \dots) \tag{2.3}$$

for solving the nonlinear equation

$$f(z) = 0, \quad f : D \subset E \to F, \tag{2.4}$$

where f is a nonlinear mapping from an open subset D of a real or complex Banach space E to a space F. To determine the convergence of Newton's method, Smale's point estimation theory presented in [167] makes use of the information of f at the initial point $z^{(0)} \in D$ instead of the domain condition in Kantorovich's theorem and leads to the important results given in Theorem 2.1. To adjust the notation with the one used in Sect. 2.5, in this section, we use parenthesized superscripts to denote the iteration index of points from the subset D, unlike Sects. 2.1–2.3 where the subscript index was employed. However, the iteration index of real sequences will be still denoted by the subscripts. In this section, we also lay the foundation prerequisite to a point estimation of the Durand–Kerner's method for the simultaneous determination of polynomial zeros, investigated in Sect. 2.5.

In 1989, X. Wang and Han [181] introduced a majorizing sequence method into the "point estimation" and obtained the following result which is more precise than Theorem 2.1.

Theorem 2.6 (Wang and Han [181]). *If* $\alpha = \alpha(z, f) \leq 3 - 2\sqrt{2}$, *then the Newton's method starting from* $z = z^{(0)}$ *is well defined, and for* $m = 0, 1, \dots$ *the inequality*

$$\| z^{(m+1)} - z^{(m)} \| \leq \frac{(1 - q^{2^m})\sqrt{(1 + \alpha)^2 - 8\alpha}}{2\alpha(1 - \eta q^{2^m - 1})(1 - \eta q^{2^m + 1} - 1)} \cdot \eta q^{2^m - 1} \| z^{(1)} - z^{(0)} \|$$

holds, where

$$\eta = \frac{1 + \alpha - \sqrt{(1 + \alpha)^2 - 8\alpha}}{1 + \alpha + \sqrt{(1 + \alpha)^2 - 8\alpha}}, \qquad q = \frac{1 - \alpha - \sqrt{(1 + \alpha)^2 - 8\alpha}}{1 - \alpha + \sqrt{(1 + \alpha)^2 - 8\alpha}}.$$

Since $3 - 2\sqrt{2} \approx 0.171573 > 0.130707$, this result obviously improves Smale's result. The constant $3 - 2\sqrt{2}$ in Theorem 2.6 is the best possible under the considered condition.

Let us note that the above two results given in Theorems 2.1 and 2.6 are based on the assumption that the sequence

$$\gamma_k = \left\| \left(f'_{(z)} \right)^{-1} f_{(z)}^{(k)} / k! \right\|, \quad k \geq 2, \tag{2.5}$$

is bounded by

$$\gamma = \sup_{k \geq 2} \gamma_k^{1/(k-1)}$$

(see Smale [167] and Sect. 2.1). However, in some concrete and special mapping, the assumption about the bound for γ_k may be suppressed. In the subsequent discussion, we present the adaptation of the work by D. Wang and Zhao [178] which offers an improvement on Smale's result. The corresponding convergence theorem gives more precise estimation under weaker condition. It is applied in Sect. 2.5 to the Durand–Kerner's method (see [32], [72], and Sect. 1.1) to state the initial conditions that depend only on starting approximations and guarantee the convergence of this method.

In what follows, we always assume that f is an analytic mapping. As in [167], let $\beta = \left\| \left(f'_{(z)} \right)^{-1} f(z) \right\|$ be the norm of the Newton's step. We consider the auxiliary functions

$$\phi(t) = \beta - \psi(t), \qquad \psi(t) = t - \sum_{k=2}^{\infty} \gamma_k t^k, \tag{2.6}$$

where γ_k is defined by (2.5), but the sequence $\{\gamma_k\}$ ($k \geq 2$) will not be assumed to be bounded by γ.

The distribution of zeros of the function $\phi(t)$ is discussed in the following lemma.

Lemma 2.6. *Let* $\max_{t>0} \psi(t)$ *exist. Then,* $\phi(t)$ *has a unique positive zero in* $[0, +\infty)$ *if* $\beta = \max_{t>0} \psi(t)$*, while* $\phi(t)$ *has exactly two different positive zeros in* $[0, +\infty)$ *if* $\beta < \max_{t>0} \psi(t)$*.*

The proof of this lemma is elementary and makes use of a simple geometrical construction and the fact that $\phi(t)$ and $\psi(t)$ are strictly convex and concave functions for $t \in (0, +\infty)$, respectively (see [178] for the complete proof).

Now, we apply Newton's method to the function $\phi(t)$ and generate a real sequence $\{t_k\}$

$$t_{m+1} = t_m - \frac{\phi(t_m)}{\phi'(t_m)} \quad (m = 0, 1, 2, \dots). \tag{2.7}$$

An important property of the sequence $\{t_m\}$ is considered in the following lemma.

Lemma 2.7. *Let t^* denote the smallest positive zero of $\phi(t)$ in $[0, +\infty)$. Then, under the condition of Lemma 2.6, the sequence $\{t_m\}$ generated by (2.7) monotonously converges to t^*.*

Proof. Assume that t' is an extremum point of $\psi(t)$. Since the functions $\phi(t)$ and $-\psi(t)$ are strictly convex, we infer that t' is their unique extremum point. Furthermore, $\phi'(t)$ is a monotonically increasing function in $[0, +\infty)$ with $\phi'(t) < 0$ in $(0, t')$, $\phi'(t) > 0$ in $(t', +\infty)$, and $\phi'(t') = 0$. Then for $t_0 < t^*$, we have $\phi(t_0) > 0$ and $\phi'(t_0) < 0$ so that from (2.7), we obtain

$$t_1 - t_0 = -\frac{\phi(t_0)}{\phi'(t_0)} > 0,$$

i.e., $t_1 > t_0$. Let us note that we may choose $t_0 = 0$ without loss of generality. Using the mean value theorem and (2.7) (for $m = 0$), we find

$$t_1 = t_0 + \frac{\phi(t^*) - \phi(t_0)}{\phi'(t_0)} = t_0 + \frac{\phi'(\xi)(t^* - t_0)}{\phi'(t_0)}, \quad \xi \in (t_0, t^*).$$

Hence

$$t_1 - t^* = (t^* - t_0)\left(\frac{\phi'(\xi)}{\phi(t_0)} - 1\right) < 0$$

because of $t^* > t_0$ and $\phi'(\xi)/\phi'(t_0) < 1$. Therefore, we have $t_1 < t^*$.

Assume now that $t_k > t_{k-1}$ and $t_k < t^*$ for $k \geq 1$. Since

$$t_{k+1} - t_k = -\frac{\phi(t_k)}{\phi'(t_k)}$$

and $\phi(t_k) > 0$, $-\phi'(t_k) > 0$, we find $t_{k+1} > t_k$. Applying the same procedure as for $k = 0$, we find

$$t_{k+1} = t_k + \frac{\phi(t^*) - \phi(t_k)}{\phi'(t_k)} = t_k + \frac{\phi'(\xi)}{\phi'(t_k)}(t^* - t_k), \quad \xi \in (t_k, t^*),$$

i.e.,

$$t_{k+1} - t^* = (t^* - t_k)\left(\frac{\phi'(\xi)}{\phi'(t_k)} - 1\right).$$

Since $t^* > t_k$ and $\phi'(\xi)/\phi'(t_k) < 1$, we get $t_{k+1} < t^*$. Therefore, we have proved by induction that the sequence $\{t_m\}$, generated by Newton's method (2.7), satisfies

$$t_{m+1} > t_m, \quad t_m < t^*, \quad \text{for all } m = 0, 1, 2, \dots.$$

Thus, the sequence $\{t_m\}$ is monotonically increasing and bounded, which means that it is convergent with the limit t^*. $\square$

According to the above consideration, we can state the following theorem.

Theorem 2.7. *Let $\{z^{(m)}\}$ be determined by the Newton's method (2.3) with the starting value $z = z^{(0)}$ and let $\{t_m\}$ be generated by (2.7). Then, under the condition of Lemma 2.6, the sequence $\{t_m\}$ is the majorizing sequence of $\{z^{(m)}\}$, i.e., the inequality*

$$\|z^{(m+1)} - z^{(m)}\| \le t_{m+1} - t_m$$

holds for all m.

Proof. In the proof of this theorem, we use the properties of $\phi(t)$ and the distribution of its zeros, given in Lemmas 2.6 and 2.7. Now, let t^* and t^{**} be the smallest and largest positive zeros of $\phi(t)$, respectively, and t' satisfies $\phi'(t') = 0$. Since

$$\phi'(t) = -1 + \sum_{k=1}^{\infty}(k+1)\gamma_{k+1}t^k,$$

we have

$$\phi'(t') = 0 \implies \sum_{k=1}^{\infty}(k+1)\gamma_{k+1}t'^k = 1.$$

According to this and having in mind that $\phi'(t)$ is a monotonically increasing function, we find

$$0 < \sum_{k=1}^{\infty}(k+1)\gamma_{k+1}t^k < 1 \quad \text{for } t \in (0, t'),$$

i.e.,

$$0 < \phi'(t) + 1 < 1.$$

Let us assume that $\|z' - z^{(0)}\| = t < t'$. Then, using the previous relations and the Taylor's expansion, we find

$$\left\|\left(f'_{(z^{(0)})}\right)^{-1}f'_{(z')} - I\right\| \le \sum_{k=1}^{\infty}\frac{\left\|\left(f'_{(z^{(0)})}\right)^{-1}f^{(k+1)}_{(z^{(0)})}\right\|}{k!}\|z' - z^{(0)}\|^k$$

$$= \sum_{k=1}^{\infty}(k+1)\gamma_{k+1}t^k = \phi'(t) + 1 < 1.$$

By using Lemma 2.1 for $c = \phi'(t) + 1$, we get

$$\left\|\left(f'_{(z')}\right)^{-1}f'_{(z^{(0)})}\right\| \le -\frac{1}{\phi'(t)}.$$

Using induction with respect to m, we shall now show that the sequence $\{z^{(m)}\}$ is well defined, and

$$\|z^{(m+1)} - z^{(m)}\| \leq t_{m+1} - t_m \quad (m = 0, 1, \ldots, t_0 = 0). \tag{2.8}$$

When $m = 0$, we obtain

$$\|z^{(1)} - z^{(0)}\| = \|\left(f'_{(z^{(0)})}\right)^{-1} f(z^{(0)})\| = \beta.$$

For $t_0 = 0$, we have $\phi(0) = \beta$ and $\phi'(0) = -1$, so that

$$t_1 - t_0 = -\frac{\phi(t_0)}{\phi'(t_0)} = \beta.$$

Thus, (2.8) holds for $m = 0$.

Let us assume that for $m \leq k - 1$, the inequalities

$$\|z^{(m+1)} - z^{(m)}\| \leq t_{m+1} - t_m \quad (m = 0, 1, \ldots, k-1)$$

are valid. Then, by summing the above inequalities, we find

$$\|z^{(k)} - z^{(0)}\| \leq \sum_{\lambda=0}^{k-1} \|z^{(\lambda+1)} - z^{(\lambda)}\| \leq \sum_{\lambda=0}^{k-1} (t_{\lambda+1} - t_\lambda) = t_k \leq t^* < t'$$

and, in the same way as above,

$$\|\left(f'_{(z^{(k)})}\right)^{-1} f'_{(z^{(0)})}\| \leq -\frac{1}{\phi'(t_k)}. \tag{2.9}$$

Now we use (2.3), (2.5), (2.7), and (2.8) and estimate

$$\|\left(f'_{(z^{(0)})}\right)^{-1} f(z^{(k)})\|$$

$$= \|\left(f'_{(z^{(0)})}\right)^{-1} \left[f(z^{(k)}) - f(z^{(k-1)}) - f'_{(z^{(k-1)})}(z^{(k)} - z^{(k-1)})\right]\|$$

$$= \left\|\int_0^1 (1-\tau)\left(f'_{(z^{(0)})}\right)^{-1} f''_{(z^{(k-1)}+\tau(z^{(k)}-z^{(k-1)}))}(z^{(k)} - z^{(k-1)})^2 d\tau\right\|$$

$$= \left\|\int_0^1 (1-\tau) \sum_{j=0}^{\infty} \frac{\left(f'_{(z^{(0)})}\right)^{-1} f^{(j+2)}_{(z^{(0)})}}{j!}\right.$$

$$\left. \times (z^{(k-1)} - z^{(0)} + \tau(z^{(k)} - z^{(k-1)}))^j (z^{(k)} - z^{(k-1)})^2 d\tau\right\|$$

$$\leq \int_0^1 (1-\tau) \sum_{j=0}^{\infty} (j+1)(j+2)\gamma_{j+2}(t_{k-1} + \tau(t_k - t_{k-1}))^j (t_k - t_{k-1})^2 d\tau$$

$$= \int_0^1 (1-\tau)\phi''(t_{k-1} + \tau(t_k - t_{k-1}))(t_k - t_{k-1})^2 d\tau$$

$$= \int_{t_{k-1}}^{t_k} (t_k - u)\phi''(u)du = \Big[(t_k - u)\phi'(u)\Big]\Big|_{t_{k-1}}^{t_k} + \int_{t_{k-1}}^{t_k} \phi'(u)du$$

$$= -(t_k - t_{k-1})\phi'(t_{k-1}) + \phi(t_k) - \phi(t_{k-1})$$

$$= \phi(t_k). \tag{2.10}$$

Finally, using (2.9) and (2.10), we get

$$\|z^{(k+1)} - z^{(k)}\| = \Big\|\big(f'_{(z^{(k)})}\big)^{-1} f(z^{(k)})\Big\|$$

$$\leq \Big\|\big(f'_{(z^{(k)})}\big)^{-1} f'(z^{(0)})\Big\| \Big\|\big(f'_{(z^{(0)})}\big)^{-1} f(z^{(k)})\Big\|$$

$$\leq -\phi(t_k)/\phi'(t_k) = t_{k+1} - t_k.$$

The proof by induction is now completed and thus, the inequality (2.8) is proved for all $m \geq 0$. $\square$

Using Theorem 2.7, one can state the convergence theorem for the Newton's method (2.3) and the existence of the solution of (2.4).

Remark 2.5. Actually, in Theorem 2.6, the Newton's method was applied to the function

$$\phi(t) = \beta - t + \frac{\gamma t^2}{1 - \gamma t}, \quad \text{where} \quad \gamma = \sup_{k \geq 2} \left\|\big(f'_{(z)}\big)^{-1} \frac{f_{(z)}^{(k)}}{k!}\right\|^{1/(k-1)}.$$

Under the condition of Theorem 2.6, X. Wang and Han [181] have proved that the sequence $\{t_m\}$ is a majorizing sequence of $\{z^{(m)}\}$. Furthermore, they have shown that the constant $\alpha_0 = 3 - 2\sqrt{2} \approx 0.172$ is the upper bound of all the best α that guarantee the convergence of the Newton's method under the point estimation conditions. Besides, the obtained estimation is optimal.

It is evident that Wang–Han's result is an improvement on Smale's result given in Theorem 2.1, while Wang–Zhao's Theorem 2.7 uses more precise estimation. In the subsequent consideration, an application of Theorem 2.7 will be demonstrated, from which one can see that the constant α_0 can be taken to be much larger than 0.172.

The subject of Theorem 2.8 is to determine a neighborhood $S(z^*, \delta)$ of the solution z^*, such that for any $z \in S(z^*, \delta)$ the condition of Theorem 2.7 is fulfilled.

Theorem 2.8. *Let*

$$\gamma^* = \sup_{k \geq 2} \left\| \left(f'_{(z^*)} \right)^{-1} f^{(k)}_{(z^*)} / k! \right\|^{1/(k-1)}$$

and let z^* be a solution of (2.4). Then, there exists a neighborhood

$$S(z^*, \delta) = \{ z \; : \; \|z - z^*\| \leq \delta, \; \gamma^* \delta \leq (3 - 2\sqrt{2})/2 \}$$

such that Theorem 2.7 is valid for any $z = z^{(0)} \in S(z^*, \delta)$.

Proof. Using the Taylor's series, we obtain

$$\left\| \left(f'_{(z^*)} \right)^{-1} f'_{(z^*)} - I \right\| \leq \left\| \sum_{k=1}^{\infty} \frac{ \left\| \left(f'_{(z^*)} \right)^{-1} f^{(k+1)}_{(z^*)} \right\| }{k!} (z - z^*) \right\|$$

$$\leq \sum_{k=1}^{\infty} (k+1)(\gamma^*)^k \delta^k = \frac{1}{(1 - \gamma^* \delta)^2} - 1 < 1,$$

where we take $z \in S(z^*, \delta)$. By Lemma 2.1, one gets

$$\left\| \left(f'_{(z)} \right)^{-1} f'_{(z^*)} \right\| \leq \frac{1}{2 - 1/(1 - \gamma^* \delta)^2} = \frac{(1 - \gamma^* \delta)^2}{h(\delta)}, \tag{2.11}$$

where $h(\delta) = 2(\gamma^*)^2 \delta^2 - 4\gamma^* \delta + 1$. Besides, we estimate

$$\left\| \left(f'_{(z^*)} \right)^{-1} f(z) \right\| \leq \left\| \sum_{k=1}^{\infty} \frac{ \left\| \left(f'_{(z^*)} \right)^{-1} f^{(k)}_{(z)} \right\| }{k!} (z - z^*)^k \right\|$$

$$\leq \sum_{k=1}^{\infty} (\gamma^*)^{k-1} \delta^k = \frac{\delta}{1 - \gamma^* \delta}. \tag{2.12}$$

Combining (2.11) and (2.12), we arrive at

$$\left\| \left(f'_{(z)} \right)^{-1} f(z) \right\| = \left\| \left(f'_{(z)} \right)^{-1} f'_{(z^*)} \right\| \left\| \left(f'_{(z^*)} \right)^{-1} f(z) \right\|$$

$$\leq \frac{\delta(1 - \gamma^* \delta)}{h(\delta)}.$$

In addition,

$$\left\| \left(f'_{(z^*)} \right)^{-1} f^{(k)}_{(z)} / k! \right\| = \left\| \sum_{\nu=1}^{\infty} \frac{ \left\| \left(f'_{(z^*)} \right)^{-1} f^{(k+\nu)}_{(z^*)} \right\| }{\nu! k!} (z - z^*)^\nu \right\|$$

$$\leq \sum_{\nu=0}^{\infty} \binom{k+\nu}{\nu} (\gamma^*)^{k+\nu-1} \delta^\nu$$

$$= (\gamma^*)^{k-1} \sum_{\nu=0}^{\infty} \binom{k+\nu}{\nu} (\gamma^*\delta)^\nu$$

$$= \frac{(\gamma^*)^{k-1}}{(1-\gamma^*\delta)^{k+1}}. \tag{2.13}$$

By virtue of (2.11) and (2.13), we obtain

$$\gamma_k = \left\| \left(f'_{(z)}\right)^{-1} f_{(z)}^{(k)}/k! \right\| \leq \frac{(\gamma^*)^{k-1}}{h(\delta)(1-\gamma^*\delta)^{k-1}}.$$

Recalling the definition of the function $\phi(t)$ (given by (2.6)) and using the previously derived estimates, we find

$$\phi(t) = \left\| \left(f'_{(z)}\right)^{-1} f(z) \right\| - t + \sum_{k=2}^{\infty} \left\| \frac{\left(f'_{(z)}\right)^{-1} f_{(z)}^{(k)}}{k!} \right\| t^k$$

$$= \frac{\delta(1-\gamma^*\delta)}{h(\delta)} - t + \sum_{k=2}^{\infty} \frac{(\gamma^*)^{k-1} t^k}{h(\delta)(1-\gamma^*\delta)^{k-1}}$$

$$= \frac{\delta(1-\gamma^*\delta)}{h(\delta)} - t + \frac{\gamma^* t^2}{h(\delta)(1-\gamma^*\delta - \gamma^* t)}$$

$$= \frac{(1-\gamma^*\delta)^2}{h(\delta)(1-\gamma^*\delta - \gamma^* t)} w(t),$$

where

$$w(t) = 2\gamma^* t^2 - (1 - 2\gamma^*\delta)t + \delta.$$

According to the above form of $\phi(t)$, we conclude that the functions $\phi(t)$ and $w(t)$ have the same zeros and $\phi(t)$ is strictly convex in $(0, +\infty)$. Under the condition

$$\delta\gamma^* < \frac{3 - 2\sqrt{2}}{2}, \tag{2.14}$$

we find

$$\Delta = (1 - 2\gamma^*\delta)^2 - 8\gamma^*\delta \geq 0, \tag{2.15}$$

which means that $w(t)$ has two positive real zeros. Therefore, $\phi(t)$ has also two positive real zeros. Having in mind Lemmas 2.6 and 2.7, this completes the proof of Theorem 2.8. □

In what follows, we apply Theorems 2.7 and 2.8 to a special algorithm for the simultaneous determination of all zeros of a polynomial.

2.5 Point Estimation of Durand–Kerner's Method

Let us consider a monic complex polynomial of degree n,

$$P(z) = z^n + a_1 z^{n-1} + \cdots + a_n = \prod_{j=1}^{n} (z - \zeta_j), \qquad (2.16)$$

where $\zeta_1, \ldots, \zeta_n$ are simple zeros of P. One of the most efficient and frequently used methods for the simultaneous approximation of all simple polynomial zeros is the Durand–Kerner's method (also known as Weierstrass' or Weierstrass–Dochev's method, see [30], [32], [72], [187])

$$z_i^{(m+1)} = z_i^{(m)} - \frac{P(z_i^{(m)})}{\displaystyle\prod_{\substack{j=1 \\ j \neq i}}^{n} (z_i^{(m)} - z_j^{(m)})} \qquad (i \in \boldsymbol{I}_n,\ m = 0, 1, \ldots\ ). \qquad (2.17)$$

Let us introduce a mapping

$$f : \mathbb{C}^n \to \mathbb{C}^n, \qquad (2.18)$$

where

$$f(z) = \begin{bmatrix} p_1(z_1, \ldots, z_n) \\ \vdots \\ p_n(z_1, \ldots, z_n) \end{bmatrix}, \qquad z = \begin{bmatrix} z_1 \\ \vdots \\ z_n \end{bmatrix} \in \mathbb{C}^n$$

and

$$p_i(z_1, \ldots, z_n) = (-1)^i \psi_i(z_1, \ldots, z_n) - a_i \quad (i = 1, \ldots, n),$$
$$(2.19)$$
$$\psi_i(z_1, \ldots, z_n) = \sum_{1 \leq j_1 < \cdots < j_i \leq n} z_{j_1} \cdots z_{j_i} \quad (i = 1, \ldots, n).$$

In 1966, Kerner [72] proved that the Durand–Kerner's method (2.17) is, actually, Newton's method (2.3) applied to the system of nonlinear equations

$$f(z) = 0, \qquad (2.20)$$

where f is defined as above. The point estimation convergence theorem of the Newton's method, presented in Sect. 2.4, will be applied in this section to the convergence analysis of the simultaneous method (2.17). First, we have to calculate the higher-order derivatives and their norms for the mapping given by (2.18) and (2.19). This requires the operation with an auxiliary polynomial defined by

$$Q(z) = z^n + b_1 z^{n-1} + \cdots + b_n = \prod_{i=1}^{n}(z - z_i), \qquad (2.21)$$

where

$$b_i = (-1)^i \psi_i(z_1, \ldots, z_n) = p_i(z_1, \ldots, z_n) + a_i \quad (i \in \mathbf{I}_n). \qquad (2.22)$$

Differentiating Q with respect to z_j $(j = 1, \ldots, n)$, we find

$$\frac{\partial Q}{\partial z_j} = -\prod_{i \neq j}(z - z_i) = \frac{\partial b_1}{\partial z_j} z^{n-1} + \cdots + \frac{\partial b_{n-1}}{\partial z_j} z + \frac{\partial b_n}{\partial z_j}.$$

Putting $z = z_i$ $(i \in \mathbf{I}_n)$ in the above relations yields

$$\begin{bmatrix} z_1^{n-1} \cdots z_1 \; 1 \\ \vdots \\ z_n^{n-1} \cdots z_n \; 1 \end{bmatrix} \begin{bmatrix} \dfrac{\partial b_1}{\partial z_1} \cdots \dfrac{\partial b_1}{\partial z_n} \\ \vdots \\ \dfrac{\partial b_n}{\partial z_1} \cdots \dfrac{\partial b_n}{\partial z_n} \end{bmatrix} = \begin{bmatrix} -Q'(z_1) & & 0 \\ & \ddots & \\ 0 & & -Q'(z_n) \end{bmatrix}. \qquad (2.23)$$

Having in mind the mapping (2.18) and (2.19), in this section, we shall use the notation that is standard in a study of systems of nonlinear equations.

From (2.22), we obtain

$$f'(z) = \begin{bmatrix} \dfrac{\partial p_1}{\partial z_1} \cdots \dfrac{\partial p_1}{\partial z_n} \\ \vdots \\ \dfrac{\partial p_n}{\partial z_1} \cdots \dfrac{\partial p_n}{\partial z_n} \end{bmatrix} = \begin{bmatrix} \dfrac{\partial b_1}{\partial z_1} \cdots \dfrac{\partial b_1}{\partial z_n} \\ \vdots \\ \dfrac{\partial b_n}{\partial z_1} \cdots \dfrac{\partial b_n}{\partial z_n} \end{bmatrix}. \qquad (2.24)$$

In view of (2.23) and (2.24), we find

$$f'(z)^{-1} = \begin{bmatrix} -\dfrac{z_1^{n-1}}{Q'(z_1)} \cdots -\dfrac{1}{Q'(z_1)} \\ \vdots \\ -\dfrac{z_n^{n-1}}{Q'(z_n)} \cdots -\dfrac{1}{Q'(z_n)} \end{bmatrix}$$

and calculate

$$A^{(k)} = f'(z)^{-1} f^{(k)}(z),$$

where $f^{(k)}(z)$ denotes the kth derivative of f. For $m_i \neq m_j$, $i \neq j$, $1 \leq i, j \leq n$, we have

$$\frac{\partial^k Q}{\partial z_{m_1} \cdots \partial z_{m_k}} = (-1)^k \prod_{k \neq m_1, \ldots, m_k}(z - z_k) = \sum_{\lambda=1}^{n} \frac{\partial^k b_\lambda}{\partial z_{m_1} \cdots \partial z_{m_k}} z^{n-\lambda},$$

while for $m_i = m_j$, we obtain

$$\frac{\partial^k Q}{\partial z_{m_1} \cdots \partial z_{m_k}} = \sum_{\lambda=1}^{n} \frac{\partial^k b_\lambda}{\partial z_{m_1} \cdots \partial z_{m_k}} z^{n-\lambda} = 0.$$

Taking into account (2.21), we find

$$A^{(k)} = f'(z)^{-1} f^{(k)}(z) = \left(-\sum_{\lambda=1}^{n} \frac{z_i^{n-\lambda}}{Q'(z_i)} \cdot \frac{\partial^k b_\lambda}{\partial z_{m_1} \cdots \partial z_{m_k}} \right)_{i, i_1, \ldots, i_k}$$

$$= \left((-1)^{k+1} \delta_{i_1 \cdots i_k} \cdot \frac{\prod_{\lambda \neq i_1, \ldots, i_k} (z_i - z_\lambda)}{Q'(z_i)} \right)_{i, i_1, \ldots, i_k},$$

where

$$\delta_{i_1, \ldots, i_k} = \begin{cases} 1, \ i_\nu = i_j, \\ 0, \ i_\nu \neq i_j. \end{cases}$$

In tensor terminology, $A^{(k)}$ is a k-order tensor.

Let T denote a k-order tensor expressed in the form $T = (t_{i, j_1, \ldots, j_k})$, where the first-order tensor is the usual matrix $T = (t_{ij})$. The norms of T are given by

$$\|T\|_1 = \max_{1 \leq j_1, \ldots, j_k \leq n} \sum_{i=1}^{n} |t_{i, j_1, \ldots, j_k}|,$$

$$\|T\|_\infty = \max_{1 \leq i \leq n} \sum_{j_1, \ldots, j_k = 1}^{n} |t_{i, j_1, \ldots, j_k}|.$$

Using these norms, two kinds of estimation of $A^{(k)}$ were found in [195]:

$$\|A^{(k)}\|_1 \leq \frac{k(n-k+1)}{n d^{k-1}}, \tag{2.25}$$

$$\|A^{(k)}\|_\infty \leq \frac{k(n-1)!}{(n-k)! d^{k-1}}, \tag{2.26}$$

where

$$d = \min_{i \neq j} |z_i - z_j|.$$

For a given point $z^{(0)} \in \mathbb{C}^n$, we denote

$$\beta_1 = \|f'(z^{(0)})^{-1} f(z^{(0)})\|_1 = \sum_{j=1}^{n} \left| \frac{P(z_j^{(0)})}{Q'(z_j^{(0)})} \right|,$$

$$\beta_\infty = \|f'(z^{(0)})^{-1} f(z^{(0)})\|_\infty = \max_{1 \leq j \leq n} \left| \frac{P(z_j^{(0)})}{Q'(z_j^{(0)})} \right|,$$

and introduce the abbreviations

$$\eta_1 = \frac{\beta_1}{d_0}, \quad \eta_\infty = \frac{\beta_\infty}{d_0}, \quad d_0 = \min_{i \neq j} |z_i^{(0)} - z_j^{(0)}|.$$

Following the construction of the scalar function $\phi(t)$ in Sect. 2.4, in a similar way, we construct the scalar functions

$$h_\infty(t) = t \left(1 + \frac{t}{d_0}\right)^{n-1} - 2t + \beta_\infty, \quad \forall t \in [0, +\infty), \qquad (2.27)$$

$$h_1(t) = \sum_{k=0}^{n-1} \frac{(n-k)t^{k+1}}{k!nd_0^k} - t + \beta_1, \quad \forall t \in [0, +\infty). \qquad (2.28)$$

Let $\tau = t/d_0$. After dividing both sides of (2.27) and (2.28) by d_0, the functions (2.27) and (2.28) become

$$\phi_\infty(\tau) = \tau(1 + \tau)^{n-1} - 2\tau + \eta_\infty,$$

$$\phi_1(\tau) = \sum_{k=1}^{n-1} \frac{(n-k)\tau^{k+1}}{k!n} - \tau + \eta_1.$$

Evidently, $\phi_1(\tau)$ and $\phi_\infty(\tau)$ possess the same properties as $h_1(t)$ and $h_\infty(t)$, respectively, which leads to the following lemma (see Zhao and D. Wang [195]).

Lemma 2.8. *The following assertions are valid:*

(i) $\phi_1(\tau)$ and $\phi_\infty(\tau)$ are strictly convex in $(0, +\infty)$.
(ii) $\phi_1(\tau)$ and $\phi_\infty(\tau)$ have real zeros in $[0, \infty)$ if and only if

$$\eta_\infty \leq -\min_{\tau > 0} \left(\tau(1 + \tau)^{n-1} - 2\tau\right), \qquad (2.29)$$

$$\eta_1 \leq -\min_{\tau > 0} \left(\sum_{k=1}^{n-1} \frac{(n-k)}{k!n}\tau^{k+1} - \tau\right). \qquad (2.30)$$

The proof of this lemma is very simple and can be found in [195]. Actually, Lemma 2.8 is, essentially, a special case of Lemma 2.6 related to the functions $\phi_\infty(\tau)$ and $\phi_1(\tau)$.

Lemma 2.9. *Assume that $\phi_\infty(\tau)$ and $\phi_1(\tau)$ satisfy (2.29) and (2.30), respectively. Let us correspond the Newton's iterations*

$$\tau_{m+1}^1 = \tau_m^1 - \frac{\phi_1(\tau_m^1)}{\phi_1'(\tau_m^1)}, \quad \tau_{m+1}^\infty = \tau_m^\infty - \frac{\phi_\infty(\tau_m^\infty)}{\phi_\infty'(\tau_m^\infty)} \quad (m = 0, 1, \ldots) \qquad (2.31)$$

to the equations

$$\phi_1(\tau) = 0, \quad \phi_\infty(\tau) = 0.$$

Then, the Newton's methods (2.31), starting from $\tau_0^1 = \tau_0^\infty = 0$, monotonously converge to the minimum roots τ_*^1 and τ_*^∞ of the above equations, respectively.

Note that this lemma is a special case of Lemma 2.7 for the functions $\phi_\infty(\tau)$ and $\phi_1(\tau)$.

Theorem 2.9. *Let us assume*

$$\eta_\infty < -\min_{\tau>0}\left(\tau(1+\tau)^{n-1} - 2\tau\right). \tag{2.32}$$

Then, the Durand–Kerner's sequence $\{z^{(m)}\}$, generated by (2.17), starting from $z = z^{(0)}$, converges to all simple roots of the equation $P(z) = 0$. Besides, we have

$$\|z^{(m)} - z^*\|_\infty \le \tau_*^\infty - \tau_m^\infty \quad (m = 0, 1, 2\ldots),$$

where τ_*^∞ is defined by (2.31) and $z^{*^T} = [\alpha_1 \cdots \alpha_n] \in \mathbb{C}^n$.

Proof. This theorem is, essentially, a special case of Theorem 2.7 for the mapping (2.18) and the function $\phi_\infty(\tau)$. In fact, according to the properties of $\phi_\infty(\tau)$ with $\tau \in [0, \tau_*^\infty]$ and (2.26), we estimate

$$\|f'(z^{(0)})^{-1}f'(z) - I\|_\infty \le \sum_{k=1}^{n-1}\left\|\frac{f'(z^{(0)})^{-1}f^{(k+1)}(z^{(0)})}{k!}\right\|_\infty \|z - z^{(0)}\|_\infty^k$$

$$\le \sum_{k=1}^{n-1}\frac{(k+1)(n-1)!}{k!(n-k-1)!}\left(\frac{t}{d_0}\right)^k \quad (t = \|z - z^{(0)}\|_\infty)$$

$$= \frac{d}{d\tau}\left(\sum_{k=1}^{n-1}\frac{(n-1)!}{k!(n-k-1)!}\tau^{k+1}\right) \quad (\tau = t/d_0)$$

$$= \phi'_\infty(\tau) + 1 < 1.$$

Hence, by Lemma 2.1 and (2.9),

$$\|f'(z)^{-1}f'(z^{(0)})\|_\infty \le -\frac{1}{\phi'_\infty(\tau)}.$$

Similarly, it can be proved

$$\|f'(z^{(0)})^{-1}f''(z)\| \le \phi''(\tau)$$

(see D. Wang and Zhao [178]). Using this result and (2.26), Zhao and D. Wang [195] have estimated

$$\|f'(z^{(0)})^{-1}f(z^{(n)})\|_\infty$$
$$\leq \|f'(z^{(0)})^{-1}[f(z^{(n)}) - f(z^{(n-1)}) - f'(z^{(n-1)})(z^{(n)} - z^{(n-1)})]\|_\infty$$
$$= \left\| \int_0^1 (1-\theta)f'(z^{(0)})^{-1}f''(z^{(n-1)} + \theta(z^{(n)} - z^{(n-1)}))(z^{(n)} - z^{(n-1)})^2 d\theta \right\|_\infty$$
$$\leq \left\| \int_0^1 (1-\theta)\phi''_\infty(\tau^\infty_{n-1} + \theta(\tau^\infty_n - \tau^\infty_{n-1}))(\tau^\infty_n - \tau^\infty_{n-1})^2 d\theta \right\|_\infty = \phi_\infty(\tau^\infty_n).$$

According to these inequalities, by induction, there follows

$$\|z^{(m+1)} - z^{(m)}\|_\infty = \|f'(z^{(m)})^{-1}f(z^{(m)})\|_\infty$$
$$\leq \|f'(z^{(m)})^{-1}f'(z^{(0)})\|_\infty \|f'(z^{(0)})^{-1}f(z^{(m)})\|_\infty$$
$$\leq -\frac{\phi_\infty(\tau^\infty_m)}{\phi'_\infty(\tau^\infty_m)} = \tau^\infty_{m+1} - \tau^\infty_m \quad (m = 0, 1, \dots). \quad \square$$

Using the properties of $\phi_1(\tau)$ and the inequalities for $\|\cdot\|_1$, Theorem 2.10 concerning the norm $\|\cdot\|_1$, similar to Theorem 2.9, can be stated.

Theorem 2.10. *Assume that*

$$\eta_1 \leq -\min_{\tau>0} \left(\sum_{k=1}^{n-1} \frac{(n-k)}{k!n}\tau^{k\,|\,1} - \tau \right). \tag{2.33}$$

Then, the Durand–Kerner's sequence $\{z^{(m)}\}$, generated by (2.17), starting from $z = z^{(0)}$, converges to all simple roots of the equation $f(z) = 0$. Besides, we have estimations

$$\|z^{(m)} - z^*\|_1 \leq \tau^1_* - \tau^1_m \quad (m = 0, 1, \dots),$$

where τ^1_m is determined by (2.31).

Conditions (2.32) and (2.33) in Theorems 2.6 and 2.7 are of limited value as far as practical problems are concerned. However, they can be suitably rearranged to the forms which are much more convenient for practical use. This is given in the following theorem, proved in [195].

Theorem 2.11. *The following inequalities are valid:*

$$-\min_{\tau>0}(\tau(1+\tau)^{n-1} - 2\tau) \geq -\min_{\tau>0} \frac{1}{n-1}(\tau e^\tau - 2\tau),$$

$$-\min_{\tau>0} \left(\sum_{k=1}^{n-1} \frac{(n-k)}{k!n}\tau^{k+1} - \tau \right) \geq -\min_{\tau>0}(\tau e^\tau - 2\tau).$$

From Theorem 2.11, it follows that if the inequalities

$$\eta_1 \leq -\min_{\tau>0}(\tau e^\tau - 2\tau), \tag{2.34}$$

$$\eta_\infty \leq -\min_{\tau>0}\frac{1}{n-1}(\tau e^\tau - 2\tau) \tag{2.35}$$

are valid, then the conditions of Theorems 2.9 and 2.10 are also valid.

The question of the existence of a neighborhood S^* of $z^{*T}=[\alpha_1 \cdots \alpha_n]$ such that Theorems 2.9 and 2.10 hold for any $z^{(0)} \in S^*$ was considered in [195].

From Wang–Han's Theorem 2.6, we see that the sequence $\{z^{(k)}\}$, generated by Newton's method, starting from $z = z^{(0)}$, converges to the solution of $f(z) = 0$ under the condition

$$\alpha = \alpha(z, f) \leq 3 - 2\sqrt{2}. \tag{2.36}$$

If Theorem 2.6 is applied to the Durand–Kerner's method with the norm $\|\cdot\|_1$, then

$$\gamma_1 = \sup_{k\geq2}\left\|\frac{A^{(k)}}{k!}\right\|_1^{1/(k-1)} = \sup_{k\geq2}\left(\frac{k(n-k+1)}{k!nd_0^{k-1}}\right)^{1/(k-1)} = \frac{(n-1)}{n}\frac{1}{d_0}.$$

Hence, by (2.36), we obtain

$$\alpha = \alpha(z^{(0)}, f) = \beta_1 \cdot \gamma_1 = \frac{(n-1)}{n}\frac{1}{d_0}\beta_1 \leq 3 - 2\sqrt{2}$$

or

$$\eta_1 \leq \frac{n}{n-1}(3 - 2\sqrt{2}), \quad \text{where } \eta_1 = \frac{\beta_1}{d_0}. \tag{2.37}$$

This convergence condition may be expressed in more explicit form as

$$\sum_{j=1}^n \left|\frac{P(z_j^{(0)})}{Q'(z_j^{(0)})}\right| \leq \frac{n}{n-1}(3 - 2\sqrt{2})d_0. \tag{2.38}$$

Taking the norm $\|\cdot\|_\infty$ yields

$$\max_{1\leq j\leq n}\left|\frac{P(z_j^{(0)})}{\prod_{\substack{k=1\\k\neq j}}^n(z_j^{(0)} - z_k^{(0)})}\right| = \max_{1\leq j\leq n}|W(z_j^{(0)})| \leq \mathrm{UB}(\eta_\infty)\,d_0, \tag{2.39}$$

where $\mathrm{UB}(a)$ denotes the upper bound of a real number a. The conditions (2.38) and (2.39) will often be encountered later in Chaps. 3–5. From (2.37), we observe that for $n = 3$, the upper bound of η_1 is $\eta_1 \leq 0.2574\ldots$, and this bound of η_1 decreases as n increases. Since $\frac{n}{n-1} \approx 1$ for sufficiently large n, it follows that for $n \geq 3$, the upper bound of η_1 will vary within the interval $[0.1716, 0.2574]$.

Let us now return to Wang–Zhao's result given in Theorem 2.9. By (2.33) and (2.34), we observe that for $n \geq 3$, the varying interval $[\underline{\eta}_1, \bar{\eta}_1]$ of the upper bound of η_1 is determined by $[\underline{\eta}_1, \bar{\eta}_1] = [0.2044, 0.3241]$, which is more precise than Wang–Han's result. Note that $\underline{\eta}_1$ does not depend on the polynomial degree n (see (2.34)). In Sect. 3.2, we will see that the results of this section are further improved.

A similar conclusion is valid for $\| \cdot \|_\infty$.

Chapter 3
Point Estimation of Simultaneous Methods ˎ

In this chapter, we are primarily interested in the construction of computationally verifiable initial conditions and the corresponding convergence analysis of the simultaneous methods presented in Sect. 1.1. These quantitative conditions predict the immediate appearance of the guaranteed and fast convergence of the considered methods. Two original procedures, based on (1) suitable localization theorems for polynomial zeros and (2) the convergence of error sequences, are applied to the most frequently used iterative methods for finding polynomial zeros.

3.1 Point Estimation and Polynomial Equations

As mentioned in Chap. 2, one of the most important problems in solving nonlinear equations is the construction of such initial conditions which provide both the guaranteed and fast convergence of the considered numerical algorithm. Smale's approach from 1981, known as "point estimation theory," examines convergence conditions in solving an equation $f(z) = 0$ using only the information of f at the initial point z_0. In the case of monic algebraic polynomials of the form

$$P(z) = z^n + a_{n-1} z^{n-1} + \cdots + a_1 z + a_0,$$

which are the main subject of our investigation in this chapter and Chaps. 4 and 5, initial conditions should be some functions of polynomial coefficients $\boldsymbol{a} = (a_0, \ldots, a_{n-1})$, its degree n, and initial approximations $\boldsymbol{z}^{(0)} = (z_1^{(0)}, \ldots, z_n^{(0)})$. A rather wide class of initial conditions can be represented by the inequality of general form

$$\phi(\boldsymbol{z}^{(0)}, \boldsymbol{a}, n) < 0. \tag{3.1}$$

M. Petković, *Point Estimation of Root Finding Methods.* Lecture Notes in Mathematics 1933,
© Springer-Verlag Berlin Heidelberg 2008

It is well known that the convergence of any iterative method for finding zeros of a given function is strongly connected with the distribution of its zeros. If these zeros are well separated, almost all algorithms show mainly good convergence properties. Conversely, in the case of very close zeros ("clusters of zeros"), almost all algorithms either fail or work with a big effort. From this short discussion, it is obvious that a measure of separation of zeros should be taken as an argument of the function ϕ given in (3.1). Since the exact zeros are unknown, we restrict ourselves to deal with the minimal distance among initial approximations $d^{(0)} = \min_{j \neq i} |z_i^{(0)} - z_j^{(0)}|$. Furthermore, the closeness of initial approximations to the wanted zeros is also an important parameter, which influences the convergence of the applied method. A measure of this closeness can be suitably expressed by a quantity of the form $h(z) = |P(z)/Q(z)|$, where $Q(z)$ does not vanish when z lies in the neighborhood $\Lambda(\zeta)$ of any zero ζ of P. For example, in the case of simple zeros of a polynomial, the choice

$$Q(z) = P'(z), \quad Q(z) = \prod_{\substack{j=1 \\ j \neq i}}^{n} (z - z_j) \quad \text{or}$$

$$|Q(z)| = |P'(z)|^{-1} \sup_{k>1} \left| \frac{P^{(k)}(z)}{k!P'(z)} \right|^{1/(k-1)} \quad (\text{see Sect. 2.1})$$

gives satisfactory results. Let us note that, considering algebraic equations, the degree of a polynomial n appears as a natural parameter in (3.1). Therefore, instead of (3.1), we can take the inequality of the form

$$\varphi(h^{(0)}, d^{(0)}, n) < 0, \tag{3.2}$$

where $h^{(0)}$ depends on P and Q at the initial point $\boldsymbol{z}^{(0)}$.

Let $\boldsymbol{I}_n := \{1, \ldots, n\}$ be the index set. For $i \in \boldsymbol{I}_n$ and $m = 0, 1, \ldots$, let us introduce the quantity

$$W_i^{(m)} = \frac{P\big(z_i^{(m)}\big)}{\displaystyle\prod_{\substack{j=1 \\ j \neq i}}^{n} \big(z_i^{(m)} - z_j^{(m)}\big)} \quad (i \in \boldsymbol{I}_n, \ m = 0, 1, \ldots), \tag{3.3}$$

which is often called Weierstrass' correction since it appeared in Weierstrass' paper [187]. In [178], D. Wang and Zhao improved Smale's result for Newton's method and applied it to the Durand–Kerner's method for the simultaneous determination of polynomial zeros (see Sect. 2.5, (2.38), and (2.39)). Their approach led in a natural way to an initial condition of the form

$$w^{(0)} \le c_n \, d^{(0)}, \tag{3.4}$$

where

$$w^{(0)} = \max_{\substack{1 \le i,j \le n \\ i \ne j}} |W_i^{(0)}|, \quad d^{(0)} = \min_{\substack{1 \le i,j \le n \\ i \ne j}} |z_i^{(0)} - z_j^{(0)}|.$$

A completely different approach presented in [112] for the same method also led to the condition of the form (3.4). In both cases, the quantity c_n was of the form $c_n = 1/(an + b)$, where a and b are suitably chosen positive constants. It turned out that initial conditions of this form are also suitable for other simultaneous methods for solving polynomial equations, as shown in the subsequent papers [5], [110], [112], [114]–[117], [119]–[121], [123], [132], [133], [136], [137], [140], [150], [151], [178], [195] and the books [20] and [118]. For these reasons, in the convergence analysis of simultaneous methods considered in this book, we will also use initial conditions of the form (3.4). We note that (3.4) is a special case of the condition (3.2). The quantity c_n, which depends only on the polynomial degree n, will be called the *inequality factor*, or the *i-factor* for brevity. We emphasize that during the last years, special attention has been paid to the increase of the *i*-factor c_n for the following obvious reason. From (3.4), we notice that a greater value of c_n allows a greater value of $|W_i^{(0)}|$. This means that cruder initial approximations can be chosen, which is of evident interest in practical realizations of numerical algorithms.

The proofs of convergence theorems of the simultaneous methods investigated in this chapter and Chaps. 4 and 5 are based on the inductive arguments. It turns out that the inequality of the form (3.4), with a specific value of c_n depending on the considered method, appears as a connecting link in the chain of inductive steps. Namely, $w^{(0)} \le c_n d^{(0)} \Rightarrow w^{(1)} \le c_n d^{(1)}$, and one may prove by induction that $w^{(0)} \le c_n d^{(0)}$ implies $w^{(m)} \le c_n d^{(m)}$ for all $m = 0, 1, 2, \ldots$.

In this chapter, we discuss the best possible values of the *i*-factor c_n appearing in the initial condition (3.4) for some efficient and frequently used iterative methods for the simultaneous determination of polynomial zeros. The reader is referred to Sect. 1.1 for the characteristics (derivation, historical notes, convergence speed) of these methods. We study the choice of "almost optimal" factor c_n. The notion "almost optimal" *i*-factor arises from (1) the presence of a system of (say) k inequalities and (2) the use of computer arithmetic of finite precision:

(1) In the convergence analysis, it is necessary to provide the validity of k substantial successive inequalities $g_1(c_n) \ge 0, \ldots, g_k(c_n) \ge 0$ (in this order), where all $g_i(c_n)$ are monotonically decreasing functions of c_n (see Fig. 3.1). The optimal value c_n should be determined as the unique solution of the corresponding equations $g_i(c_n) = 0$. Unfortunately, all equations cannot be satisfied simultaneously and we are constrained to find such c_n which makes the inequalities $g_i(c_n) \ge 0$ as sharp as possible. Since $g_i(c_n) \ge 0$ succeeds $g_j(c_n) \ge 0$ for $j < i$, we first find c_n so that the inequality $g_1(c_n) \ge 0$ is as sharp as possible and check the validity

of all remaining inequalities $g_2(c_n) \geq 0, \ldots, g_k(c_n) \geq 0$. If some of them are not valid, we decrease c_n and repeat the process until all inequalities are satisfied. For demonstration, we give a particular example on Fig. 3.1. The third inequality $g_3(c_n) \geq 0$ is not satisfied for $c_n^{(1)}$, so that c_n takes a smaller value $c_n^{(2)}$ satisfying all three inequalities. In practice, the choice of c_n is performed iteratively, using a programming package, in our book *Mathematica* 6.0.

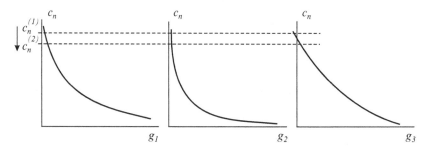

Fig. 3.1 The choice of i-factor c_n iteratively

(2) Since computer arithmetic of finite precision is employed, the optimal value (the exact solution of $g_i(c_n) = 0$, if it exists for some i) cannot be represented exactly, so that c_n should be decreased for a few bits to satisfy the inequalities $g_i(c_n) > 0$. The required conditions (in the form of inequalities $g_i(c_n) \geq 0$) are still satisfied with great accuracy. We stress that this slight decrease of the i-factor c_n with respect to the optimal value is negligible from a practical point of view. For this reason, the constants a and b appearing in $c_n = 1/(an + b)$ are rounded for all methods considered in this book.

The entries of c_n, obtained in this way and presented in this chapter, are increased (and, thus, improved) compared with those given in the literature, which means that newly established initial conditions for the guaranteed convergence of the considered methods are weakened (see Fig. 3.3).

We note that all considerations in this book are given for $n \geq 3$, taking into account that algebraic equations of the order ≤ 2 are trivial and their numerical treatment is unnecessary. In our analysis, we will sometimes omit the iteration index m, and new entries in the later $(m + 1)$th iteration will be additionally stressed by the symbol $\hat{\ }$ ("hat"). For example, instead of

$$z_i^{(m)}, z_i^{(m+1)}, W_i^{(m)}, W_i^{(m+1)}, d^{(m)}, d^{(m+1)}, N_i^{(m)}, N_i^{(m+1)}, \quad \text{etc.,}$$

we will write

$$z_i, \hat{z}_i, W_i, \widehat{W}_i, d, \hat{d}, N_i, \widehat{N}_i.$$

According to this, we denote

$$w = \max_{1 \le i \le n} |W_i|, \quad \widehat{w} = \max_{1 \le i \le n} |\widehat{W}_i|.$$

This denotation will also be used in the subsequent study in Chaps. 4 and 5.

3.2 Guaranteed Convergence: Correction Approach

In this chapter, we present two procedures in the study of the guaranteed convergence of simultaneous methods (1) the approach based on iterative corrections and (2) the approach based on convergent sequences. Both schemes will be applied to the most frequently used simultaneous zero-finding methods in considerable details.

Applying the first method (1), we will deal with a real function $t \mapsto g(t)$ defined on $(0, 1)$ by

$$g(t) = \begin{cases} 1 + 2t, & 0 < t \le \dfrac{1}{2} \\ \dfrac{1}{1-t}, & \dfrac{1}{2} < t < 1 \end{cases}$$

The minorizing function of $g(t)$ on $(0, 1)$ is given in the following lemma whose proof is elementary.

Lemma 3.1. *Let*

$$s_m(t) = \sum_{i=0}^{m} t^i + t^m \quad (t \in (0, 1), \ m = 1, 2, \ldots).$$

Then, $s_m(t) < g(t)$.

Most of the iterative methods for the simultaneous determination of simple zeros of a polynomial can be expressed in the form

$$z_i^{(m+1)} = z_i^{(m)} - C_i\big(z_1^{(m)}, \ldots, z_n^{(m)}\big) \quad (i \in \boldsymbol{I}_n, \ m = 0, 1, \ldots), \tag{3.5}$$

where $z_1^{(m)}, \ldots, z_n^{(m)}$ are some distinct approximations to simple zeros $\zeta_1, \ldots, \zeta_n$, respectively, obtained in the mth iterative step by the method (3.5). In what follows, the term

$$C_i^{(m)} = C_i\big(z_1^{(m)}, \ldots, z_n^{(m)}\big) \quad (i \in \boldsymbol{I}_n)$$

will be called the *iterative correction* or simply the *correction*.

Let $\Lambda(\zeta_i)$ be a sufficiently close neighborhood of the zero ζ_i $(i \in I_n)$. In this book, we consider a class of iterative methods of the form (3.5) with corrections C_i which can be expressed as

$$C_i(z_1, \ldots, z_n) = \frac{P(z_i)}{F_i(z_1, \ldots, z_n)} \quad (i \in I_n), \tag{3.6}$$

where the function $(z_1, \ldots, z_n) \mapsto F_i(z_1, \ldots, z_n)$ satisfies the following conditions for each $i \in I_n$ and distinct approximations $z_1, \ldots, z_n$:

1° $F_i(\zeta_1, \ldots, \zeta_n) \neq 0$,

2° $F_i(z_1, \ldots, z_n) \neq 0$ for any $(z_1, \ldots, z_n) \in \Lambda(\zeta_1) \times \cdots \times \Lambda(\zeta_n) =: Y$,

3° $F_i(z_1, \ldots, z_n)$ is continuous in $\mathbb{C}^n$.

Starting from mutually disjoint approximations $z_1^{(0)}, \ldots, z_n^{(0)}$, the iterative method (3.5) produces n sequences of approximations $\{z_i^{(m)}\}$ $(i \in I_n)$ which, under certain convenient conditions, converge to the polynomial zeros. Indeed, if we find the limit values

$$\lim_{m \to \infty} z_i^{(m)} = \zeta_i \quad (i \in I_n),$$

then having in mind (3.6) and the conditions 1°–3°, we obtain from (3.5)

$$0 = \lim_{m \to \infty} \left(z_i^{(m)} - z_i^{(m+1)} \right) = \lim_{m \to \infty} C_i \left(z_1^{(m)}, \ldots, z_n^{(m)} \right)$$

$$= \lim_{m \to \infty} \frac{P\left(z_i^{(m)}\right)}{F_i\left(z_1^{(m)}, \ldots, z_n^{(m)}\right)} = \frac{P(\zeta_i)}{F_i(\zeta_1, \ldots, \zeta_n)} \quad (i \in I_n).$$

Hence $P(\zeta_i) = 0$, i.e., ζ_i is a zero of the polynomial P.

Theorem 3.1 has the key role in our convergence analysis of simultaneous methods presented in this section and Chap. 4 (see M. Petković, Carstensen, and Trajković [112]).

Theorem 3.1. *Let the iterative method* (3.5) *have the iterative correction of the form* (3.6) *for which the conditions* 1°–3° *hold, and let* $z_1^{(0)}, \ldots, z_n^{(0)}$ *be distinct initial approximations to the zeros of* P. *If there exists a real number* $\beta \in (0, 1)$ *such that the following two inequalities*

(i) $\left| C_i^{(m+1)} \right| \leq \beta \left| C_i^{(m)} \right|$ $(m = 0, 1, \ldots)$,

(ii) $\left| z_i^{(0)} - z_j^{(0)} \right| > g(\beta) \left(\left| C_i^{(0)} \right| + \left| C_j^{(0)} \right| \right)$ $(i \neq j, \ i, j \in I_n)$

are valid, then the iterative method (3.5) *is convergent.*

Proof. Let us define disks $D_i^{(m)} := \{z_i^{(m+1)}; |C_i^{(m)}|\}$ for $i \in I_n$ and $m = 0, 1, \ldots$, where $z_i^{(m+1)}$ and $C_i^{(m)}$ are approximations and corrections appearing in (3.5). Then for a fixed $i \in I_n$, we have

$$D_i^{(m)} = \{z_i^{(m)} - C_i^{(m)}; |C_i^{(m)}|\} = \{z_i^{(m-1)} - C_i^{(m-1)} - C_i^{(m)}; |C_i^{(m)}|\} = \cdots$$
$$= \{z_i^{(0)} - C_i^{(0)} - C_i^{(1)} - \cdots - C_i^{(m)}; |C_i^{(m)}|\} \subset \{z_i^{(0)}; r_i^{(m)}\},$$

where
$$r_i^{(m)} = |C_i^{(0)}| + \cdots + |C_i^{(m-1)}| + 2|C_i^{(m)}|.$$

Using (i), we find $|C_i^{(k)}| \le \beta^k |C_i^{(0)}|$ $(k = 1, 2, \ldots, \beta < 1)$ so that, according to Lemma 3.1 and the definition of the function $g(t)$,

$$r_i^{(m)} \le |C_i^{(0)}|(1 + \beta + \cdots + \beta^m + \beta^m) < g(\beta)|C_i^{(0)}|.$$

Therefore, for each $i \in I_n$, we have the inclusion

$$D_i^{(m)} \subset S_i := \{z_i^{(0)}; g(\beta)|C_i^{(0)}|\},$$

which means that the disk S_i contains all the disks $D_i^{(m)}$ $(m = 0, 1, \ldots)$. In regard to this and the definition of disks $D_i^{(m)}$, we can illustrate the described situation by Fig. 3.2.

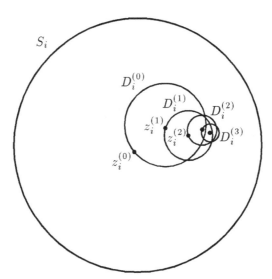

Fig. 3.2 Inclusion disk S_i contains all disks $D_i^{(m)}$

The sequence $\{z_i^{(m)}\}$ of the centers of the disks $D_i^{(m)}$ forms a Cauchy's sequence in the disk $S_i \supset D_i^{(m)}$ $(m = 0, 1, \ldots)$. Since the metric subspace S_i

is complete (as a closed set in $\mathbb{C}$), there exists a unique point $z_i^* \in S_i$ such that

$$z_i^{(m)} \to z_i^* \quad \text{as} \quad m \to \infty \quad \text{and} \quad z_i^* \in S_i.$$

Since

$$z_i^{(m+1)} = z_i^{(m)} - \frac{P(z_i^{(m)})}{F(z_1^{(m)}, \ldots, z_n^{(m)})}$$

and $F(z_1^{(m)}, \ldots, z_n^{(m)})$ does not vanish whenever $(z_1, \ldots, z_n) \in Y$, there follows

$$|P(z_i^{(m)})| = |F(z_1^{(m)}, \ldots, z_n^{(m)})(z_i^{(m+1)} - z_i^{(m)})|$$
$$\leq |F(z_1^{(m)}, \ldots, z_n^{(m)})| \, |z_i^{(m+1)} - z_i^{(m)}|.$$

Taking the limit when $m \to \infty$, we obtain

$$|P(z_i^*)| \leq \lim_{m \to \infty} |F(z_1^{(m)}, \ldots, z_n^{(m)})| \lim_{m \to \infty} |z_i^{(m+1)} - z_i^{(m)}| = 0,$$

which means that the limit points $z_1^*, \ldots, z_n^*$ of the sequences $\{z_1^{(m)}\}, \ldots, \{z_n^{(m)}\}$ are, actually, the zeros of the polynomial P. To complete the proof of the theorem, it is necessary to show that each of the sequences $\{z_i^{(m)}\}$ ($i \in I_n$) converges to one and only one zero of P. Since $z_i^{(m)} \in S_i$ for each $i \in I_n$ and $m = 0, 1, \ldots$, it suffices to prove that the disks $S_1, \ldots, S_n$ are mutually disjoint, i.e. (according to (1.67)),

$$|z_i^{(0)} - z_j^{(0)}| = |\text{mid } S_i - \text{mid } S_j| > \text{rad } S_i + \text{rad } S_j = g(\beta)\big(|C_i^{(0)}| + |C_j^{(0)}|\big) \quad (i \neq j),$$

which reduces to (ii). □

In this section and Chap. 4, we will apply Theorem 3.1 to some iterative methods for the simultaneous approximation of simple zeros of a polynomial. We will assume that an iterative method is well defined if $F(z_1, \ldots, z_n) \neq 0$ under the stated initial conditions and for each array of approximations $(z_1, \ldots, z_n)$ obtained in the course of the iterative procedure.

The convergence analysis of simultaneous methods considered in this section is essentially based on Theorem 3.1 and the four relations connecting the quantities $|W_i|$ (Weierstrass' corrections), d (minimal distance between approximations), and $|C_i|$ (iterative corrections). These relations are referred to as W–D, W–W, C–C, and C–W inequalities according to the quantities involved, and read thus:

(W–D): $w^{(0)} \lesssim c_n d^{(0)},$ (3.7)

(W–W): $\left|W_i^{(m+1)}\right| \leq \delta_n \left|W_i^{(m)}\right| \quad (i \in I_n, \ m = 0, 1, \ldots),$ (3.8)

(C–C): $\qquad |C_i^{(m+1)}| \leq \beta_n |C_i^{(m)}| \qquad (i \in \boldsymbol{I}_n,\ m = 0, 1, \ldots),$ (3.9)

(C–W): $\qquad |C_i^{(m)}| \leq \lambda_n \dfrac{|W_i^{(m)}|}{c_n} \qquad (i \in \boldsymbol{I}_n,\ m = 0, 1, \ldots).$ (3.10)

Here, c_n, δ_n, β_n, and λ_n are real positive constants depending only on the polynomial degree n. The W–D inequality (3.7) defines the initial condition for the guaranteed convergence of an iterative method and plays the main role in the convergence analysis based on the relations (3.7)–(3.10).

The convergence analysis consists of two steps:

1° Starting from the W–D inequality (3.7), derive the W–W inequality (3.8) for each $m = 0, 1, \ldots$. The i-factor c_n has to be chosen so that $\delta_n < 1$ holds. In this way, the convergence of the sequences of Weierstrass' corrections $\{W_i^{(m)}\}$ $(i \in \boldsymbol{I}_n)$ to 0 is ensured.

2° Derive the C–C inequality (3.9) for each $m = 0, 1, \ldots$ under the condition (3.7). The choice of the i-factor c_n must provide the validity of the C–W inequality (3.10) and the inequalities

$$\beta_n < 1 \tag{3.11}$$

and

$$\lambda_n < \frac{1}{2g(\beta_n)}. \tag{3.12}$$

The last requirement arises from the following consideration. Assume that (3.7) implies the inequality (3.10) for all $i \in \boldsymbol{I}_n$. Then using (3.7), we obtain

$$\left| z_i^{(0)} - z_j^{(0)} \right| \geq d^{(0)} \geq \frac{w^{(0)}}{c_n} \geq \frac{|C_i^{(0)}| + |C_j^{(0)}|}{2\lambda_n}.$$

Hence, to provide the inequality (ii) in Theorem 3.1, it is necessary to be $1/(2\lambda_n) > g(\beta_n)$ (the inequality (3.12)) where, according to the conditions of Theorem 3.1, the (positive) argument β_n must be less than 1 (the inequality (3.11)). Note that the requirement $\beta_n < 1$ is also necessary to ensure the contraction of the correction terms (see (3.9)) and, thus, the convergence of the considered simultaneous method.

In the subsequent analysis, we will apply the described procedure to some favorable simultaneous methods. This procedure requires certain bounds of the same type and, to avoid the repetition, we give them in the following lemma.

Lemma 3.2. *For distinct complex numbers* $z_1, \ldots, z_n$ *and* $\hat{z}_1, \ldots, \hat{z}_n$, *let*

$$d = \min_{\substack{1 \leq i, j \leq n \\ i \neq j}} |z_i - z_j|, \quad \hat{d} = \min_{\substack{1 \leq i, j \leq n \\ i \neq j}} |\hat{z}_i - \hat{z}_j| \quad (i \in \boldsymbol{I}_n).$$

If

$$|\hat{z}_i - z_i| \le \lambda_n d \quad (i \in \boldsymbol{I}_n, \, \lambda_n < 1/2), \tag{3.13}$$

then

$$|\hat{z}_i - z_j| \ge (1 - \lambda_n)d \quad (i \in \boldsymbol{I}_n), \tag{3.14}$$

$$|\hat{z}_i - \hat{z}_j| \ge (1 - 2\lambda_n)d \quad (i \in \boldsymbol{I}_n), \tag{3.15}$$

and

$$\left| \prod_{j \ne i} \frac{\hat{z}_i - z_j}{\hat{z}_i - \hat{z}_j} \right| \le \left(1 + \frac{\lambda_n}{1 - 2\lambda_n} \right)^{n-1}. \tag{3.16}$$

Proof. Applying the triangle inequality, we find

$$|\hat{z}_i - z_j| \ge |z_i - z_j| - |\hat{z}_i - z_i| \ge d - \lambda_n d = (1 - \lambda_n)d$$

and

$$|\hat{z}_i - \hat{z}_j| \ge |z_i - z_j| - |\hat{z}_i - z_i| - |\hat{z}_j - z_j| \ge d - \lambda_n d - \lambda_n d = (1 - 2\lambda_n)d. \tag{3.17}$$

From

$$\prod_{j \ne i} \frac{\hat{z}_i - z_j}{\hat{z}_i - \hat{z}_j} = \prod_{j \ne i} \left(1 + \frac{\hat{z}_j - z_j}{\hat{z}_i - \hat{z}_j} \right)$$

and

$$\left| \frac{\hat{z}_j - z_j}{\hat{z}_i - \hat{z}_j} \right| \le \frac{\lambda_n d}{(1 - 2\lambda_n)d} = \frac{\lambda_n}{1 - 2\lambda_n},$$

we obtain

$$\left| \prod_{j \ne i} \frac{\hat{z}_i - z_j}{\hat{z}_i - \hat{z}_j} \right| = \prod_{j \ne i} \left| 1 + \frac{\hat{z}_j - z_j}{\hat{z}_i - \hat{z}_j} \right| \le \prod_{j \ne i} \left(1 + \left| \frac{\hat{z}_j - z_j}{\hat{z}_i - \hat{z}_j} \right| \right)$$

$$\le \prod_{j \ne i} \left(1 + \frac{\lambda_n}{1 - 2\lambda_n} \right) = \left(1 + \frac{\lambda_n}{1 - 2\lambda_n} \right)^{n-1}. \quad \square$$

Remark 3.1. Since $\hat{d} \le |\hat{z}_i - \hat{z}_j|$, from (3.17) we obtain

$$\hat{d} \le (1 - 2\lambda_n)d. \tag{3.18}$$

In what follows, we apply Theorem 3.1 to the convergence analysis of four frequently used simultaneous zero-finding methods.

The Durand–Kerner's Method

One of the most frequently used iterative methods for the simultaneous determination of simple zeros of a polynomial is the Durand–Kerner's (or Weierstrass') method defined by

$$z_i^{(m+1)} = z_i^{(m)} - W_i^{(m)} \quad (i \in \boldsymbol{I}_n, \ m = 0, 1, \ldots), \tag{3.19}$$

where $W_i^{(m)}$ is given by (3.3). In this case, the iterative correction term is equal to Weierstrass' correction, i.e., $C_i = W_i = P(z_i)/F_i(z_1, \ldots, z_n)$, where

$$F_i(z_1, \ldots, z_n) = \prod_{\substack{j=1 \\ j \neq i}}^{n} (z_i - z_j) \quad (i \in \boldsymbol{I}_n).$$

To simplify the denotation, we will omit sometimes the iteration index m in the sequel and denote quantities in the subsequent $(m+1)$th iteration by $\hat{\ }$ ("hat"). It will be always assumed that the polynomial degree n is not smaller than 3.

Lemma 3.3. *Let $z_1, \ldots, z_n$ be distinct approximations and let*

$$w \leq c_n d, \tag{3.20}$$

$$c_n \in (0, 0.5), \tag{3.21}$$

$$\delta_n := \frac{(n-1)\, c_n}{1 - c_n} \left(1 + \frac{c_n}{1 - 2c_n} \right)^{n-1} \leq 1 - 2c_n \tag{3.22}$$

hold. Then:

(i) $\left| \widehat{W}_i \right| \leq \delta_n \left| W_i \right|$.

(ii) $\hat{w} \leq c_n \hat{d}$.

Proof. Let $\lambda_n = c_n$. From (3.19) and (3.20), there follows

$$|\hat{z}_i - z_i| = |W_i| \leq w \leq c_n d. \tag{3.23}$$

According to this and Lemma 3.2, we obtain

$$|\hat{z}_i - z_j| \geq (1 - c_n) d \tag{3.24}$$

and

$$|\hat{z}_i - \hat{z}_j| \geq (1 - 2c_n) d. \tag{3.25}$$

From the iterative formula (3.19), it follows

$$\frac{W_i}{\hat{z}_i - z_i} = -1,$$

so that

$$\sum_{j=1}^{n} \frac{W_j}{\hat{z}_i - z_j} + 1 = \frac{W_i}{\hat{z}_i - z_i} + \sum_{j \neq i} \frac{W_j}{\hat{z}_i - z_j} + 1 = \sum_{j \neq i} \frac{W_j}{\hat{z}_i - z_j}. \tag{3.26}$$

Putting $z = \hat{z}_i$ in the polynomial representation by Lagrange's interpolation formula

$$P(z) = \left(\sum_{j=1}^{n} \frac{W_j}{z - z_j} + 1 \right) \prod_{j=1}^{n} (z - z_j), \tag{3.27}$$

we find by (3.26)

$$P(\hat{z}_i) = (\hat{z}_i - z_i) \left(\sum_{j \neq i} \frac{W_j}{\hat{z}_i - z_j} \right) \prod_{j \neq i} (\hat{z}_i - z_j).$$

After dividing with $\prod_{j \neq i} (\hat{z}_i - \hat{z}_j)$, one obtains

$$\widehat{W}_i = \frac{P(\hat{z}_i)}{\prod\limits_{j \neq i} (\hat{z}_i - \hat{z}_j)} = (\hat{z}_i - z_i) \left(\sum_{j \neq i} \frac{W_j}{\hat{z}_i - z_j} \right) \prod_{j \neq i} \left(1 + \frac{\hat{z}_j - z_j}{\hat{z}_i - \hat{z}_j} \right). \tag{3.28}$$

Using the inequalities (3.20), (3.23)–(3.25) and Lemma 3.2, from (3.28), we estimate

$$|\widehat{W}_i| \leq |\hat{z}_i - z_i| \sum_{j \neq i} \frac{|W_j|}{|\hat{z}_i - z_j|} \prod_{j \neq i} \left(1 + \frac{|\hat{z}_j - z_j|}{|\hat{z}_i - \hat{z}_j|} \right)$$

$$\leq |W_i| \frac{(n-1)w}{(1 - c_n)d} \left(1 + \frac{c_n d}{(1 - 2c_n)d} \right)^{n-1}$$

$$\leq |W_i| \frac{(n-1)c_n}{1 - c_n} \left(1 + \frac{c_n}{1 - 2c_n} \right)^{n-1}$$

$$= \delta_n |W_i|.$$

This proves the assertion (i) of the lemma.

Since

$$\hat{d} = \min_{\substack{1 \leq i, j \leq n \\ i \neq j}} |\hat{z}_i - \hat{z}_j|,$$

from (3.25), one obtains

$$\hat{d} \geq (1 - 2\lambda_n)d = (1 - 2c_n)d, \quad \text{i.e.,} \quad d \leq \frac{\hat{d}}{1 - 2c_n}.$$

According to the last inequality and (3.22), we estimate

$$|\widehat{W}_i| \leq \delta_n |W_i| \leq \delta_n c_n d \leq \frac{\delta_n}{1 - 2c_n} c_n \hat{d} \leq c_n \hat{d}.$$

Therefore, the assertion (ii) holds. $\square$

Theorem 3.2. *Let the assumptions from Lemma 3.3 hold. If $z_1^{(0)}, \ldots, z_n^{(0)}$ are distinct approximations for which the initial condition*

$$w^{(0)} \le c_n d^{(0)} \tag{3.29}$$

is valid, then the Durand–Kerner's method (3.19) is convergent.

Proof. It is sufficient to prove the assertions (i) and (ii) of Theorem 3.1 taking $C_i^{(m)} = W_i^{(m)}$ in this particular case.

According to (ii) of Lemma 3.3, we conclude that (3.29) provides the implication $w^{(0)} \le c_n d^{(0)} \Rightarrow w^{(1)} \le c_n d^{(1)}$. In a similar way, we show the implication

$$w^{(m)} \le c_n d^{(m)} \implies w^{(m+1)} \le c_n d^{(m+1)},$$

proving by induction that the initial condition (3.29) implies the inequality

$$w^{(m)} \le c_n d^{(m)} \tag{3.30}$$

for each $m = 1, 2, \ldots$. Hence, by (i) of Lemma 3.3, we get

$$\left| W_i^{(m+1)} \right| \le \delta_n \left| W_i^{(m)} \right| = \beta_n \left| W_i^{(m)} \right| \tag{3.31}$$

for each $m = 0, 1, \ldots$. Let us note that (3.31) is the W–W inequality of the form (3.8), but also the C–C inequality of the form (3.9) since $C_i = W_i$ in this particular case with $\beta_n = \delta_n$, where δ_n is given by (3.22). Therefore, the assertion (i) holds true.

In a similar way as for (3.25), under the condition (3.29), we prove the inequality

$$\left| z_i^{(m+1)} - z_j^{(m+1)} \right| \ge (1 - 2c_n) d^{(m)} > 0 \quad (i \ne j, \ i, j \in \boldsymbol{I}_n, \ m = 0, 1, \ldots),$$

so that

$$F_i\big(z_1^{(m)}, \ldots, z_n^{(m)}\big) = \prod_{i \ne j} (z_i^{(m)} - z_j^{(m)}) \ne 0$$

in each iteration. Therefore, the Durand–Kerner's method (3.19) is well defined.

Since $\beta_n = \delta_n$, from (3.22), we see that $\beta_n < 1$ (necessary condition (3.11)), and the function g is well defined. To prove (ii) of Theorem 3.1, we have to show that the inequality (3.12) is valid. If $\beta_n \ge 1/2$, then (3.12) becomes

$$\frac{1}{1 - \beta_n} < \frac{1}{2\lambda_n},$$

which is equivalent to (3.22). If $\beta_n < 1/2$, then (3.12) reduces to

$$1 + \beta_n < \frac{1}{2\lambda_n}, \quad \text{i.e.,} \quad \lambda_n = c_n < \frac{1}{2(1 + 2\beta_n)} \in (0.25, 0.5),$$

which holds according to the assumption (3.21) of Lemma 3.3. Since we have proved both assertions (i) and (ii) of Theorem 3.1, we conclude that the Durand–Kerner's method (3.19) is convergent. □

The choice of the "almost optimal" value of c_n is considered in the following lemma.

Lemma 3.4. *The i-factor c_n given by*

$$c_n = \frac{1}{An + B}, \quad A = 1.76325, \quad B = 0.8689425, \tag{3.32}$$

satisfies the conditions (3.21) *and* (3.22).

Proof. Since $c_n \leq c_3 \approx 0.16238$, it follows that $c_n \in (0, 0.5)$ and (3.21) holds true.

To prove (3.22), it is sufficient to prove the inequality

$$\eta_n := \frac{\delta_n}{1 - 2c_n} = \frac{n-1}{1 - c_n} \frac{c_n}{1 - 2c_n} \left(1 + \frac{c_n}{1 - 2c_n}\right)^{n-1} \leq 1. \tag{3.33}$$

Since

$$\lim_{n \to \infty} \frac{1}{1 - c_n} - 1, \quad \lim_{n \to \infty} \left(1 + \frac{c_n}{1 - 2c_n}\right)^{\frac{1 - 2c_n}{c_n}} = e, \quad \lim_{n \to \infty} \frac{(n-1)c_n}{1 - 2c_n} = \frac{1}{A},$$

where $A = 1.76325$ appears in (3.32), we obtain

$$\lim_{n \to \infty} \eta_n = \frac{1}{A} e^{1/A} < 0.99998 < 1.$$

Since the sequence $\{\eta_n\}$, defined by (3.33), is monotonically increasing for $n \geq 3$, we have $\eta_n < \eta_\infty < 0.99998 < 1$. □

Remark 3.2. The constant $A = 1.76325$ is determined as the reciprocal value of the approximate solution of the equation $xe^x = 1$, and chosen so that it satisfies the inequality $e^{1/A}/A < 1$ (to fulfill the condition $\lim_{n \to \infty} \eta_n < 1$). The use of an approximate solution of the equation $xe^x = 1$ instead of the exact solution (that cannot be represented in floating-point arithmetic of finite precision) just leads to the notion of the "almost optimal" i-factor. Taking a greater number of decimal digits for A (and, consequently, for B, see Remark 3.3), we can make the inequality (3.34) arbitrarily sharp. In this way, we can improve the i-factor c_n to the desired (optimal) extent but, from a practical point of view, such improvement is negligible.

Remark 3.3. Note that the coefficient B in (3.32), not only for the Durand–Kerner's method but also for other methods, is chosen so that the entries δ_n, β_n, and λ_n appearing in the W–W, C–C, and C–W inequalities (3.8)–(3.10) ensure the validity of these inequalities for a particular n, most frequently for $n = 3$. For example, this coefficient for the Durand–Kerner's method is $B = 0.8689425$.

According to Theorem 3.1 and Lemma 3.4, we can state the convergence theorem, which considers initial conditions for the guaranteed convergence of the Durand–Kerner's method.

Theorem 3.3. *The Durand–Kerner's method is convergent under the condition*

$$w^{(0)} < \frac{d^{(0)}}{1.76325n + 0.8689425}. \tag{3.34}$$

Remark 3.4. The sign $<$ ("strongly less") in the inequality (3.34) differs from " $\leq$ " used in the previous consideration since the concrete choice of A and B in (3.14) yields $\delta_n < 1 - 2c_n$ in (3.22) (also "strongly less"). This is also the case in all remaining methods presented in this book, so that the subsequent situations of this type will not be explained again.

Some authors have considered initial conditions in the form of the inequality

$$\| \boldsymbol{W}^{(0)} \|_1 = \sum_{i=1}^{n} |W_i^{(0)}| \leq \Omega_n d^{(0)}, \quad \boldsymbol{W}^{(0)} = (W_1^{(0)}, \ldots, W_n^{(0)}),$$

instead of the condition (3.7). Obviously, one can take $\Omega_n = n \, c_n$ since (3.29) implies

$$|W_i^{(0)}| \leq c_n d^{(0)} \quad (i = 1, \ldots, n).$$

As already mentioned, the choice of c_n and Ω_n as large as possible permits cruder initial approximations.

We recall some previous ranges concerned with the bounds of Ω_n for $n \geq 3$. X. Wang and Han obtained in [181]

$$\Omega_n = \frac{n}{n-1} \left(3 - 2\sqrt{2} \right) \in (0.1716, 0.2574) \quad (n \geq 3).$$

D. Wang and Zhao improved in [178] the above result yielding the interval

$$\Omega_n \in (0.2044, 0.3241) \quad (n \geq 3).$$

Batra [5] and M. Petković et al. [120] have dealt with $c_n = 1/(2n)$, which gives $\Omega_n = 0.5$. The choice of c_n in this section (see (3.32)) yields

$$\Omega_3 = 3c_3 = 0.48712$$

and

$$\Omega_n \in \left(4c_4, \lim_{n \to \infty} nc_n \right) = \left(0.50493, \frac{1}{A} \right) = (0.50493, 0.56713) \quad (n \geq 4),$$

which improves all previous results.

The Börsch-Supan's Method

Börsch-Supan's third-order method for the simultaneous approximations of all simple zeros of a polynomial, presented for the first time in [10] and later in [95], is defined by the iterative formula

$$z_i^{(m+1)} = z_i^{(m)} - \frac{W_i^{(m)}}{1 + \sum\limits_{\substack{j=1 \\ j \neq i}}^{n} \dfrac{W_j^{(m)}}{z_i^{(m)} - z_j^{(m)}}} \quad (i \in \boldsymbol{I}_n,\ m = 0, 1, \ldots), \quad (3.35)$$

where $W_i^{(m)}$ are given by (3.3) (see Sect. 1.1). This formula has the form (3.5) with the correction

$$C_i(z_1, \ldots, z_n) = \frac{P(z_i)}{F_i(z_1, \ldots, z_n)} \quad (i \in \boldsymbol{I}_n),$$

where

$$F_i(z_1, \ldots, z_n) = \left(1 + \sum_{j \neq i} \frac{W_j}{z_i - z_j} \right) \prod_{j \neq i} (z_i - z_j) \quad (i \in \boldsymbol{I}_n).$$

Before establishing the main convergence theorems, we prove two auxiliary results.

Lemma 3.5. *Let $z_1, \ldots, z_n$ be distinct complex numbers and let*

$$c_n \in \left(0, \frac{1}{n+1} \right) \tag{3.36}$$

and

$$w \leq c_n d. \tag{3.37}$$

Then:

(i) $\dfrac{c_n}{\lambda_n} \leq \left| 1 + \sum\limits_{j \neq i} \dfrac{W_j}{z_i - z_j} \right| \leq 2 - \dfrac{c_n}{\lambda_n}.$

(ii) $|\hat{z}_i - z_i| \leq \dfrac{\lambda_n}{c_n} |W_i| \leq \lambda_n d.$

(iii) $|\hat{z}_i - z_j| \geq (1 - \lambda_n) d.$

(iv) $|\hat{z}_i - \hat{z}_j| \geq (1 - 2\lambda_n) d.$

(v) $\left| \sum\limits_{j=1}^{n} \dfrac{W_j}{\hat{z}_i - z_j} + 1 \right| \leq \dfrac{(n-1)\lambda_n c_n}{1 - \lambda_n}.$

(vi) $\left| \prod\limits_{j \neq i} \dfrac{\hat{z}_i - z_j}{\hat{z}_i - \hat{z}_j} \right| \leq \left(1 + \dfrac{\lambda_n}{1 - 2\lambda_n} \right)^{n-1},$

where $\lambda_n = \dfrac{c_n}{1 - (n-1)c_n}.$

Proof. Since $1 - 2\lambda_n = \dfrac{1 - (n+1)c_n}{1 - (n-1)c_n}$, from (3.36), it follows $0 < 1 - 2\lambda_n < 1$, hence $\lambda_n \in (0, 0.5)$. By (3.37) and the definition of d, we obtain

$$\left| 1 + \sum_{j \neq i} \frac{W_j}{z_i - z_j} \right| \geq 1 - \sum_{j \neq i} \frac{|W_j|}{|z_i - z_j|} \geq 1 - \frac{(n-1)w}{d} \geq 1 - (n-1)\,c_n = \frac{c_n}{\lambda_n}$$

and

$$\left| 1 + \sum_{j \neq i} \frac{W_j}{z_i - z_j} \right| \leq 1 + \frac{(n-1)w}{d} \leq 1 + (n-1)\,c_n = 2 - \frac{c_n}{\lambda_n},$$

which proves (i). By (i) and (3.37), we prove (ii):

$$|\hat{z}_i - z_i| = \frac{|W_i|}{\left| 1 + \displaystyle\sum_{j \neq i} \frac{W_j}{z_i - z_j} \right|} \leq \frac{|W_i|}{1 - (n-1)\,c_n} = \frac{\lambda_n}{c_n}|W_i| \leq \lambda_n d.$$

The assertions (iii), (iv), and (vi) follow directly according to Lemma 3.2. Omitting the iteration index, from (3.35), we find

$$\frac{W_i}{\hat{z}_i - z_i} = -1 - \sum_{j \neq i} \frac{W_j}{z_i - z_j},$$

so that

$$\left| \sum_{j=1}^{n} \frac{W_j}{\hat{z}_i - z_j} + 1 \right| = \left| \frac{W_i}{\hat{z}_i - z_i} + \sum_{j \neq i} \frac{W_j}{\hat{z}_i - z_j} + 1 \right| = \left| \sum_{j \neq i} \frac{W_j(z_i - \hat{z}_i)}{(\hat{z}_i - z_j)(z_i - z_j)} \right|.$$

Hence, using (3.37), (ii), and (iii), it follows

$$\left| \sum_{j=1}^{n} \frac{W_j}{\hat{z}_i - z_j} + 1 \right| \leq |\hat{z}_i - z_i| \sum_{j \neq i} \frac{|W_j|}{|\hat{z}_i - z_j||z_i - z_j|} \leq \frac{(n-1)\lambda_n c_n}{1 - \lambda_n},$$

which means that (v) is also true. This completes the proof of the lemma. $\square$

According to Lemma 3.5, we can prove the following assertions.

Lemma 3.6. *Let $z_1, \ldots, z_n$ be distinct approximations and let the assumptions (3.36) and (3.37) of Lemma 3.5 hold. In addition, let*

$$\delta_n := \frac{(n-1)\lambda_n^2}{1 - \lambda_n}\left(1 + \frac{\lambda_n}{1 - 2\lambda_n} \right)^{n-1} \leq 1 - 2\lambda_n \tag{3.38}$$

be valid. Then:

(i) $|\widehat{W}_i| \leq \delta_n |W_i|$.

(ii) $\widehat{w} \leq c_n \widehat{d}$.

Proof. Setting $z = \hat{z}_i$ in (3.27), where $\hat{z}_i$ is a new approximation produced by the Börsch-Supan's method (3.35), we obtain

$$P(\hat{z}_i) = (\hat{z}_i - z_i) \left(\sum_{j=1}^{n} \frac{W_j}{\hat{z}_i - z_j} + 1 \right) \prod_{j \neq i} (\hat{z}_i - z_j).$$

After dividing with $\prod_{j \neq i} (\hat{z}_i - \hat{z}_j)$, we get

$$\widehat{W}_i = (\hat{z}_i - z_i) \left(\sum_{j=1}^{n} \frac{W_j}{\hat{z}_i - z_j} + 1 \right) \prod_{j \neq i} \frac{\hat{z}_i - z_j}{\hat{z}_i - \hat{z}_j}.$$

Using the bounds (ii), (v), and (vi) of Lemma 3.5, we estimate

$$|\widehat{W}_i| = |\hat{z}_i - z_i| \left| \sum_{j=1}^{n} \frac{W_j}{\hat{z}_i - z_j} + 1 \right| \left| \prod_{j \neq i} \frac{\hat{z}_i - z_j}{\hat{z}_i - \hat{z}_j} \right|$$

$$\leq |W_i| \frac{(n-1)\lambda_n^2}{1 - \lambda_n} \left(1 + \frac{\lambda_n}{1 - 2\lambda_n} \right)^{n-1},$$

i.e., $|\widehat{W}_i| \leq \delta_n |W_i|$. Therefore, the assertion (i) holds true.

According to (iv) of Lemma 3.5, there follows

$$\hat{d} \geq (1 - 2\lambda_n)d.$$

This inequality, together with (i) of Lemma 3.6 and (3.38), gives (ii), i.e.,

$$|\widehat{W}_i| \leq \delta_n |W_i| \leq \frac{\delta_n}{1 - 2\lambda_n} c_n \hat{d} \leq c_n \hat{d}. \quad \square$$

Theorem 3.4. *Let the assumptions from Lemmas 3.5 and 3.6 hold and, in addition, let*

$$\beta_n := \left(\frac{2\lambda_n}{c_n} - 1 \right) \delta_n < 1 \tag{3.39}$$

and

$$g(\beta_n) < \frac{1}{2\lambda_n}. \tag{3.40}$$

If $z_1^{(0)}, \ldots, z_n^{(0)}$ are distinct initial approximations satisfying

$$w^{(0)} \leq c_n d^{(0)}, \tag{3.41}$$

then the Börsch-Supan's method (3.35) *is convergent.*

Proof. It is sufficient to prove (i) and (ii) of Theorem 3.1 for the iterative correction given by

$$C_i^{(m)} = \frac{W_i^{(m)}}{1 + \sum_{j \neq i} \dfrac{W_j^{(m)}}{z_i^{(m)} - z_j^{(m)}}} \quad (i \in I_n, \ m = 0, 1, \dots)$$

(see (3.35)). By virtue of Lemma 3.6, which holds under the conditions (3.36), (3.38), and (3.41), we can prove by induction that

$$w^{(m+1)} \leq \delta_n w^{(m)} \leq \frac{\delta_n}{1 - 2\lambda_n} c_n d^{(m+1)} \leq c_n d^{(m+1)}$$

holds for each $m = 0, 1, \dots$.

Starting from the assertion (i) of Lemma 3.5, under the condition (3.41), we prove by induction

$$F_i\left(z_1^{(m)}, \dots, z_n^{(m)}\right) = \left(1 + \sum_{j \neq i} \frac{W_j^{(m)}}{z_i^{(m)} - z_j^{(m)}}\right) \prod_{j \neq i} \left(z_i^{(m)} - z_j^{(m)}\right) \neq 0$$

for each $i \in I_n$ and $m = 0, 1, \dots$. Therefore, the Börsch-Supan's method (3.35) is well defined in each iteration.

Using (i) of Lemma 3.5, we find

$$|C_i| = \frac{|W_i|}{\left|1 + \sum_{j \neq i} \dfrac{W_j}{z_i - z_j}\right|} \leq \frac{\lambda_n}{c_n} |W_i|, \qquad (3.42)$$

so that for the next iterative step we obtain by Lemma 3.5 and (i) of Lemma 3.6

$$|\widehat{C_i}| \leq \frac{\lambda_n}{c_n} |\widehat{W_i}| \leq \frac{\lambda_n \delta_n}{c_n} \frac{|W_i|}{\left|1 + \sum_{j \neq i} \dfrac{W_j}{z_i - z_j}\right|} \left|1 + \sum_{j \neq i} \frac{W_j}{z_i - z_j}\right|$$

$$= \frac{\lambda_n \delta_n}{c_n} |C_i| \left|1 + \sum_{j \neq i} \frac{W_j}{z_i - z_j}\right| \leq \frac{\lambda_n \delta_n}{c_n} \left(2 - \frac{c_n}{\lambda_n}\right) |C_i|$$

$$= \delta_n \left(\frac{2\lambda_n}{c_n} - 1\right) |C_i| = \beta_n |C_i|,$$

where $\beta_n < 1$ (the assumption (3.39)). Using the same argumentation, we prove by induction

$$\left|C_i^{(m+1)}\right| \leq \beta_n \left|C_i^{(m)}\right|$$

for each $i \in I_n$ and $m = 0, 1, \ldots$.

By (3.41) and (3.42), we estimate

$$\frac{1}{\lambda_n}|C_i^{(0)}| \leq \frac{|W_i^{(0)}|}{c_n} \leq d^{(0)}.$$

According to this and (3.40), we see that

$$|z_i^{(0)} - z_j^{(0)}| \geq d^{(0)} \geq \frac{w^{(0)}}{c_n} \geq \frac{1}{2\lambda_n}\left(|C_i^{(0)}| + |C_j^{(0)}|\right)$$
$$> g(\beta_n)\left(|C_i^{(0)}| + |C_j^{(0)}|\right)$$

holds for each $i \neq j$, $i, j \in I_n$. This proves (ii) of Theorem 3.1. The validity of (i) and (ii) of Theorem 3.1 shows that the Börsch-Supan's method (3.35) is convergent under the given conditions. □

The choice of the i-factor c_n is considered in the following lemma.

Lemma 3.7. *The i-factor c_n defined by*

$$c_n = \begin{cases} \dfrac{1}{n + \frac{9}{2}}, & n = 3, 4 \\[4mm] \dfrac{1}{\frac{309}{200}n + 5}, & n \geq 5 \end{cases} \tag{3.43}$$

satisfies the condition of Theorem 3.4.

The proof of this lemma is elementary and it is derived by a simple analysis of the sequences $\{\beta_n\}$ and $\{g(\beta_n)\}$.

According to Lemma 3.7 and Theorem 3.4, we may state the following theorem.

Theorem 3.5. *The Börsch-Supan's method (3.35) is convergent under the condition (3.41), where c_n is given by (3.43).*

Tanabe's Method

In Sect. 1.1, we have presented the third-order method, often referred to as Tanabe's method

$$z_i^{(m+1)} = z_i^{(m)} - W_i^{(m)}\left(1 - \sum_{j \neq i} \frac{W_j^{(m)}}{z_i^{(m)} - z_j^{(m)}}\right) \quad (i \in I_n, \ m = 0, 1, \ldots). \tag{3.44}$$

As in the previous cases, before stating initial conditions that ensure the guaranteed convergence of the method (3.44), we give first some necessary bounds using the previously introduced notation and omitting the iteration index for simplicity.

Lemma 3.8. *Let* $z_1, \ldots, z_n$ *be distinct approximations and let*

$$c_n \in \left(0, \frac{1}{1 + \sqrt{2n-1}}\right). \tag{3.45}$$

If the inequality

$$w \leq c_n d \tag{3.46}$$

holds, then for $i, j \in \mathbf{I}_n$ *we have:*

(i) $\dfrac{\lambda_n}{c_n} = 1 + (n-1)c_n \geq \left|1 - \displaystyle\sum_{j \neq i} \dfrac{W_j}{z_i - z_j}\right| \geq 1 - (n-1)c_n = 2 - \dfrac{\lambda_n}{c_n}.$

(ii) $|\hat{z}_i - z_i| \leq \dfrac{\lambda_n}{c_n}|W_i| \leq \lambda_n d.$

(iii) $|\hat{z}_i - z_j| \geq (1 - \lambda_n)d.$

(iv) $|\hat{z}_i - \hat{z}_j| \geq (1 - 2\lambda_n)d.$

(v) $\left|\displaystyle\sum_{j=1}^{n} \dfrac{W_j}{\hat{z}_i - z_j} + 1\right| \leq \dfrac{(n-1)c_n^2}{(2c_n - \lambda_n)(1 - \lambda_n)}\left(\lambda_n + (n-1)c_n\right).$

(vi) $\displaystyle\prod_{j \neq i} \left|\dfrac{\hat{z}_i - z_j}{\hat{z}_i - \hat{z}_j}\right| \leq \left(1 + \dfrac{\lambda_n}{1 - 2\lambda_n}\right)^{n-1},$

where $\lambda_n = (1 + (n-1)c_n)c_n.$

Proof. We omit the proofs of the assertions (i)–(iv) and (vi) since they are quite similar to those given in Lemma 3.5. To prove (v), we first introduce

$$\sigma_i = \sum_{j \neq i} \frac{W_j}{z_i - z_j}.$$

Then

$$|\sigma_i| \leq \frac{(n-1)w}{d} \leq (n-1)c_n \quad \text{and} \quad \frac{|\sigma_i|}{1 - |\sigma_i|} \leq \frac{(n-1)c_n}{1 - (n-1)c_n}. \tag{3.47}$$

From the iterative formula (3.44), we obtain

$$\frac{W_i}{\hat{z}_i - z_i} = -\frac{1}{1 - \displaystyle\sum_{j \neq i} \dfrac{W_j}{z_i - z_j}},$$

so that by (3.47) it follows

$$
\left| \sum_{j=1}^{n} \frac{W_j}{\hat{z}_i - z_j} + 1 \right| = \left| \frac{W_i}{\hat{z}_i - z_i} + \sum_{j \neq i} \frac{W_j}{\hat{z}_i - z_j} + 1 \right| = \left| 1 - \frac{1}{1 - \sigma_i} + \sum_{j \neq i} \frac{W_j}{\hat{z}_i - z_j} \right|
$$

$$
= \frac{1}{|1 - \sigma_i|} \left| \sum_{j \neq i} \frac{W_j}{\hat{z}_i - z_j} - \sum_{j \neq i} \frac{W_j}{z_i - z_j} - \sigma_i \sum_{j \neq i} \frac{W_j}{z_i - z_j} \right|
$$

$$
\leq \frac{1}{|1 - \sigma_i|} \left| \sum_{j \neq i} \frac{W_j (z_i - \hat{z}_i)}{(\hat{z}_i - z_j)(z_i - z_j)} \right| + \frac{|\sigma_i|}{1 - |\sigma_i|} \left| \sum_{j \neq i} \frac{W_j}{\hat{z}_i - z_j} \right|
$$

$$
\leq \frac{1}{1 - (n-1)c_n} |z_i - \hat{z}_i| \sum_{j \neq i} \frac{|W_j|}{|\hat{z}_i - z_j||z_i - z_j|}
$$

$$
+ \frac{(n-1)c_n}{1 - (n-1)c_n} \sum_{j \neq i} \frac{|W_j|}{|\hat{z}_i - z_j|}.
$$

Hence, by (ii), (iii), (3.46), and (3.47), we estimate

$$
\left| \sum_{j=1}^{n} \frac{W_j}{\hat{z}_i - z_j} + 1 \right| \leq \frac{\lambda d}{1 - (n-1)c_n} \frac{(n-1)w}{(1 - \lambda_n)d \cdot d} + \frac{(n-1)c_n}{1 - (n-1)c_n} \frac{(n-1)w}{(1 - \lambda_n)d}
$$

$$
= \frac{(n-1)c_n^2}{(2c_n - \lambda_n)(1 - \lambda_n)} \Big(\lambda_n + (n-1)c_n \Big). \quad \square
$$

Remark 3.5. The inequalities (iv) and (vi) require $2\lambda_n < 1$ or $2c_n(1+(n-1)c_n) < 1$. This inequality will be satisfied if $c_n < 1/(1 + \sqrt{2n-1})$, which is true according to (3.45).

Lemma 3.9. *Let $z_1, \ldots, z_n$ be distinct approximations and let the assumptions (3.45) and (3.46) of Lemma 3.8 hold. If the inequality*

$$
\delta_n := \frac{(n-1)c_n \lambda_n}{(2c_n - \lambda_n)(1 - \lambda_n)} \Big(\lambda_n + (n-1)c_n \Big) \Big(1 + \frac{\lambda_n}{1 - 2\lambda_n} \Big)^{n-1} \leq 1 - 2\lambda_n \quad (3.48)
$$

is valid, then:

(i) $|\widehat{W}_i| \leq \delta_n |W_i|$.
(ii) $\hat{w} \leq c_n \hat{d}$.

Proof. Putting $z = \hat{z}_i$ in (3.27), where $\hat{z}_i$ is a new approximation obtained by Tanabe's method (3.44), and dividing with $\prod_{j \neq i} (\hat{z}_i - \hat{z}_j)$, we obtain

$$
\widehat{W}_i = \frac{P(\hat{z}_i)}{\prod_{j \neq i} (\hat{z}_i - \hat{z}_j)} = (\hat{z}_i - z_i) \left[\sum_{j=1}^{n} \frac{W_j}{\hat{z}_i - z_j} + 1 \right] \prod_{j \neq i} \frac{\hat{z}_i - z_j}{\hat{z}_i - \hat{z}_j}.
$$

From the last relation, we obtain by (ii), (v), and (vi) of Lemma 3.8

$$
\begin{aligned}
|\widehat{W}_i| &= |\hat{z}_i - z_i| \left| \sum_{j=1}^{n} \frac{W_j}{\hat{z}_i - z_j} + 1 \right| \prod_{j \neq i} \left| \frac{\hat{z}_i - z_j}{\hat{z}_i - \hat{z}_j} \right| \\
&\leq \frac{\lambda_n}{c_n} |W_i| \frac{(n-1)c_n^2}{(2c_n - \lambda_n)(1 - \lambda_n)} \left(\lambda_n + (n-1)c_n \right) \left(1 + \frac{\lambda_n}{1 - 2\lambda_n} \right)^{n-1} \\
&= \delta_n |W_i|,
\end{aligned}
$$

which proves (i).

Using (iv) of Lemma 3.8, we find

$$
\hat{d} \geq (1 - 2\lambda_n)d.
$$

Combining this inequality with (i) of Lemma 3.9, (3.45), and (3.48), we prove (ii):

$$
|\widehat{W}_i| \leq \delta_n |W_i| \leq \delta_n c_n d \leq \frac{\delta_n}{1 - 2\lambda_n} c_n \hat{d} \leq c_n \hat{d}. \quad \square
$$

Theorem 3.6. *Let the assumptions of Lemmas 3.7 and 3.8 be valid and let*

$$
\beta_n := \frac{\lambda_n \delta_n}{2c_n - \lambda_n} < 1 \tag{3.49}
$$

and

$$
g(\beta_n) < \frac{1}{2\lambda_n}. \tag{3.50}
$$

If $z_1^{(0)}, \ldots, z_n^{(0)}$ are distinct initial approximations satisfying

$$
w^{(0)} \leq c_n d^{(0)}, \tag{3.51}
$$

then the Tanabe's method (3.44) is convergent.

Proof. In Lemma 3.9 (assertion (ii)), we derived the implication $w \leq c_n d \Rightarrow \hat{w} \leq c_n \hat{d}$. Using a similar procedure, we prove by induction that the initial condition (3.51) implies the inequality $w^{(m)} \leq c_n d^{(m)}$ for each $m = 1, 2, \ldots$. Therefore, all assertions of Lemmas 3.8 and 3.9 are valid for each $m = 1, 2, \ldots$. For example, we have

$$
|W_i^{(m+1)}| \leq \delta_n |W_i^{(m)}| \quad (i \in \boldsymbol{I}_n, \ m = 0, 1, \ldots). \tag{3.52}
$$

From the iterative formula (3.44), we see that corrections $C_i^{(m)}$ are given by

$$
C_i^{(m)} = W_i^{(m)} \left(1 - \sum_{j \neq i} \frac{W_j^{(m)}}{z_i^{(m)} - z_j^{(m)}} \right) \quad (i \in \boldsymbol{I}_n). \tag{3.53}
$$

This correction has the required form

$$C_i^{(m)} = P(z_i^{(m)})/F(z_1^{(m)}, \ldots, z_n^{(m)}),$$

where

$$F_i(z_1^{(m)}, \ldots, z_n^{(m)}) = \frac{\prod\limits_{j \neq i}(z_i^{(m)} - z_j^{(m)})}{1 - \sum\limits_{j \neq i} \dfrac{W_j^{(m)}}{z_i^{(m)} - z_j^{(m)}}} \qquad (i \in \boldsymbol{I}_n).$$

According to (i) of Lemma 3.8, it follows $F_i(z_1^{(m)}, \ldots, z_n^{(m)}) \neq 0$, which means that the Tanabe's method is well defined in each iteration.

We now prove the first part of the theorem which is concerned with the monotonicity of the sequences $\{C_i^{(m)}\}$ ($i \in \boldsymbol{I}_n$) of corrections. Starting from (3.53) and omitting iteration indices, we find by (ii) of Lemma 3.8 (which is valid under the condition (3.51))

$$|C_i| = |W_i| \left| 1 - \sum_{j \neq i} \frac{W_j}{z_i - z_j} \right| \leq \frac{\lambda_n}{c_n} |W_i|. \qquad (3.54)$$

According to (3.52)–(3.54) and by the inequalities (i) of Lemma 3.8, we obtain

$$|\widehat{C}_i| \leq \frac{\lambda_n}{c_n} |\widehat{W}_i| \leq \frac{\lambda_n \delta_n}{c_n} |W_i| = \frac{\lambda_n \delta_n}{c_n} \cdot \frac{\left| W_i \left(1 - \sum\limits_{j \neq i} \dfrac{W_j}{z_i - z_j} \right) \right|}{\left| 1 - \sum\limits_{j \neq i} \dfrac{W_j}{z_i - z_j} \right|}$$

$$\leq \frac{\lambda_n \delta_n}{c_n(2 - \lambda_n/c_n)} |C_i| = \beta_n |C_i|,$$

where $\beta_n < 1$ (assumption (3.49)). By induction, it is proved that the inequality $|C_i^{(m+1)}| \leq \beta_n |C_i^{(m)}|$ holds for each $i = 1, \ldots, n$ and $m = 0, 1, \ldots$.

By (3.51) and (3.54), we estimate

$$\frac{1}{\lambda_n} |C_i^{(0)}| \leq \frac{w^{(0)}}{c_n} \leq d^{(0)}.$$

According to this, (3.50), and (3.51), we conclude that

$$|z_i^{(0)} - z_j^{(0)}| \geq d^{(0)} \geq \frac{w^{(0)}}{c_n} \geq \frac{1}{2\lambda_n} (|C_i^{(0)}| + |C_j^{(0)}|) > g(\beta_n)(|C_i^{(0)}| + |C_j^{(0)}|)$$

holds for each $i \neq j$, $i, j \in \boldsymbol{I}_n$. This proves (ii) of Theorem 3.1. □

Lemma 3.10. *The i-factor c_n given by $c_n = 1/(3n)$ satisfies the condition (3.45), (3.48), (3.49), and (3.50).*

Proof. Obviously, $c_n = 1/(3n) < 1/(1+\sqrt{2n-1})$. Furthermore, the sequence $\{\delta_n\}$ is monotonically increasing so that

$$\delta_n < \lim_{n \to +\infty} \delta_n = \frac{2}{9}e^{4/9} \approx 0.3465 < 0.35 \quad \text{for every } n \geq 3.$$

We adopt $\delta_n = 0.35$ and prove that (3.48) holds; indeed,

$$\delta_n = 0.35 < 1 - 2\lambda_n = 1 - \frac{2(4n-1)}{9n^2} \quad \left(\geq \frac{59}{81} \approx 0.728\right)$$

for every $n \geq 3$.

For $\delta_n = 0.35$ and $c_n = 1/(3n)$, the sequence $\{\beta_n\}$ defined by

$$\beta_n = \frac{\delta_n \lambda_n}{2c_n - \lambda_n} = \frac{0.35(4n-1)}{2n+1}$$

is monotonically increasing so that

$$\beta_n < \lim_{n \to +\infty} \delta_n = 0.7 < 1 \quad (n \geq 3),$$

which means that (3.49) is valid.

Finally, we check the validity of the inequality (3.50) taking $\beta_n = 0.7$. We obtain

$$g(\beta_n) = g(0.7) = \frac{1}{1-0.7} = 3.333... < \frac{1}{2\lambda_n} = \frac{9n^2}{2(4n-1)} \quad \left(\geq \frac{81}{22} \approx 3.68\right),$$

wherefrom we conclude that the inequality (3.50) holds for every $n \geq 3$. □

According to Lemma 3.10 and Theorem 3.1, the following theorem is stated.

Theorem 3.7. *The Tanabe's method (3.44) is convergent under condition (3.51), where $c_n = 1/(3n)$.*

The Chebyshev-Like Method

In Sect. 1.1, we have presented the iterative fourth-order method of Chebyshev's type

$$z_i^{(m+1)} = z_i^{(m)} - \frac{W_i^{(m)}}{1 + G_{1,i}^{(m)}}\left(1 - \frac{W_i^{(m)} G_{2,i}^{(m)}}{(1 + G_{1,i}^{(m)})^2}\right) \quad (i \in \boldsymbol{I}_n, \, m = 0, 1, 2, \ldots),$$

$$(3.55)$$

proposed by M. Petković, Tričković, and Đ. Herceg [146]. Before stating initial
conditions that guarantee the convergence of the method (3.55), three lemmas
which concern some necessary bounds and estimations are given first.

Lemma 3.11. *Let $z_1, \ldots, z_n$ be distinct approximations to the zeros $\zeta_1, \ldots, \zeta_n$
of a polynomial P and let $\hat{z}_1, \ldots, \hat{z}_n$ be new approximations obtained by the
iterative formula (3.55). If the inequality*

$$w < c_n d, \quad c_n = \frac{2}{5n+3} \quad (n \geq 3) \tag{3.56}$$

holds, then for all $i \in I_n$ we have:

(i) $\dfrac{3n+5}{5n+3} < |1 + G_{1,i}| < \dfrac{7n+1}{5n+3}.$

(ii) $|\hat{z}_i - z_i| \leq \dfrac{\lambda_n}{c_n}|W_i| \leq \lambda_n d,$ *where* $\lambda_n = \dfrac{2(9n^2 + 34n + 21)}{(3n+5)^3}.$

Proof. According to the definition of the minimal distance d and the inequality
(3.56), it follows

$$|G_{1,i}| \leq \sum_{j \neq i} \frac{|W_j|}{|z_i - z_j|} < (n-1)c_n, \quad |G_{2,i}| \leq \sum_{j \neq i} \frac{|W_j|}{|z_i - z_j|^2} < \frac{(n-1)c_n}{d},$$

$$\tag{3.57}$$

so that we find

$$|1 + G_{1,i}| \geq 1 - \sum_{j \neq i} \frac{|W_j|}{|z_i - z_j|} > 1 - (n-1)c_n = \frac{3n+5}{5n+3}$$

and

$$|1 + G_{1,i}| \leq 1 + \sum_{j \neq i} \frac{|W_j|}{|z_i - z_j|} < 1 + (n-1)c_n = \frac{7n+1}{5n+3}.$$

Therefore, the assertion (i) of Lemma 3.11 is proved.

Using (i) and (3.57), we estimate

$$\left| \frac{W_i}{1 + G_{1,i}} \right| < \frac{w}{1 - (n-1)c_n} < \frac{2}{3n+5}d \tag{3.58}$$

and

$$\left| \frac{W_i G_{2,i}}{(1 + G_{1,i})^2} \right| < \frac{w(n-1)c_n/d}{(1 - (n-1)c_n)^2} < \frac{c_n^2(n-1)}{(1 - (n-1)c_n)^2} \leq \frac{4(n-1)}{(3n+5)^2}. \tag{3.59}$$

Using (3.58) and (3.59), we obtain the bound (ii):

$$
\begin{aligned}
|\hat{z}_i - z_i| = |C_i| &= \left| \frac{W_i}{1 + G_{1,i}} \left(1 - \frac{W_i G_{2,i}}{(1 + G_{1,i})^2} \right) \right| \\
&\leq \frac{|W_i|}{|1 + G_{1,i}|} \left(1 + \frac{|W_i G_{2,i}|}{|1 + G_{1,i}|^2} \right) \\
&< |W_i| \cdot \frac{5n + 3}{3n + 5} \left(1 + \frac{4(n - 1)}{(3n + 5)^2} \right) \\
&< \frac{2(9n^2 + 34n + 21)}{(3n + 5)^3} d = \lambda_n d. \quad \square
\end{aligned}
$$

According to Lemma 3.2 and the assertion (ii) of Lemma 3.11, under the condition (3.56), we have

$$
|\hat{z}_i - z_j| > (1 - \lambda_n)d \quad (i, j \in I_n), \tag{3.60}
$$

$$
|\hat{z}_i - \hat{z}_j| > (1 - 2\lambda_n)d \quad (i, j \in I_n), \tag{3.61}
$$

and

$$
\left| \prod_{j \neq i} \frac{\hat{z}_i - z_j}{\hat{z}_i - \hat{z}_j} \right| < \left(1 + \frac{\lambda_n}{1 - 2\lambda_n} \right)^{n-1}. \tag{3.62}
$$

Let us note that the necessary condition $\lambda_n < 1/2$ is satisfied under the condition (3.56).

Remark 3.6. Since (3.61) is valid for arbitrary pair $i, j \in I_n$ and $\lambda_n < 1/2$ if (3.56) holds, there follows

$$
\hat{d} = \min_{j \neq i} |\hat{z}_i - \hat{z}_j| > (1 - 2\lambda_n)d. \tag{3.63}
$$

Lemma 3.12. If the inequality (3.56) holds, then

(i) $|\widehat{W}_i| < 0.22|W_i|$.

(ii) $\widehat{w} < \dfrac{2}{5n + 3}\hat{d}$.

Proof. For distinct points $z_1, \ldots, z_n$, we use the polynomial representation (3.27) and putting $z = \hat{z}_i$ in (3.27), we find

$$
P(\hat{z}_i) = (\hat{z}_i - z_i) \left(\frac{W_i}{\hat{z}_i - z_i} + \sum_{j \neq i} \frac{W_j}{\hat{z}_i - z_j} + 1 \right) \prod_{j \neq i} (\hat{z}_i - z_j).
$$

After division with $\prod\limits_{j\neq i}(\hat{z}_i - \hat{z}_j)$, we get

$$\widehat{W}_i = \frac{P(\hat{z}_i)}{\prod\limits_{j\neq i}(\hat{z}_i - \hat{z}_j)} = (\hat{z}_i - z_i)\left(\frac{W_i}{\hat{z}_i - z_i} + \sum_{j\neq i}\frac{W_j}{\hat{z}_i - z_j} + 1\right)\prod_{j\neq i}\left(\frac{\hat{z}_i - z_j}{\hat{z}_i - \hat{z}_j}\right).$$

(3.64)

From the iterative formula (3.55), it follows

$$\frac{W_i}{\hat{z}_i - z_i} = \frac{-(1 + G_{1,i})}{1 - \dfrac{W_i G_{2,i}}{(1 + G_{1,i})^2}} = -1 - \frac{G_{1,i} + \dfrac{W_i G_{2,i}}{(1 + G_{1,i})^2}}{1 - \dfrac{W_i G_{2,i}}{(1 + G_{1,i})^2}}.$$

(3.65)

Then

$$\frac{W_i}{\hat{z}_i - z_i} + \sum_{j\neq i}\frac{W_j}{\hat{z}_i - z_j} + 1 = -1 - \frac{G_{1,i} + \dfrac{W_i G_{2,i}}{(1 + G_{1,i})^2}}{1 - \dfrac{W_i G_{2,i}}{(1 + G_{1,i})^2}} + \sum_{j\neq i}\frac{W_j}{\hat{z}_i - z_j} + 1$$

$$= \frac{-\sum\limits_{j\neq i}\dfrac{W_j}{z_i - z_j} + \sum\limits_{j\neq i}\dfrac{W_j}{\hat{z}_i - z_j} - \dfrac{W_i G_{2,i}}{(1 + G_{1,i})^2} - \dfrac{W_i G_{2,i}}{(1 + G_{1,i})^2}\sum\limits_{j\neq i}\dfrac{W_j}{\hat{z}_i - z_j}}{1 - \dfrac{W_i G_{2,i}}{(1 + G_{1,i})^2}}$$

$$= \frac{-(\hat{z}_i - z_i)\sum\limits_{j\neq i}\dfrac{W_j}{(z_i - z_j)(\hat{z}_i - z_j)} - \dfrac{W_i G_{2,i}}{(1 + G_{1,i})^2}\left(1 + \sum\limits_{j\neq i}\dfrac{W_j}{\hat{z}_i - z_j}\right)}{1 - \dfrac{W_i G_{2,i}}{(1 + G_{1,i})^2}}.$$

From the last formula, we obtain by (3.59), (3.60), the definition of the minimal distance, and (ii) of Lemma 3.11

$$\left|\frac{W_i}{\hat{z}_i - z_i} + \sum_{j\neq i}\frac{W_j}{\hat{z}_i - z_j} + 1\right|$$

$$\leq \frac{|\hat{z}_i - z_i|\sum\limits_{j\neq i}\dfrac{|W_j|}{|z_i - z_j||\hat{z}_i - z_j|} + \left|\dfrac{W_i G_{2,i}}{(1 + G_{1,i})^2}\right|\left(1 + \sum\limits_{j\neq i}\dfrac{|W_j|}{|\hat{z}_i - z_j|}\right)}{1 - \left|\dfrac{W_i G_{2,i}}{(1 + G_{1,i})^2}\right|}$$

$$< \frac{8(135n^5 + 594n^4 + 646n^3 - 292n^2 - 821n - 262)}{(5n + 3)(9n^2 + 26n + 29)(27n^3 + 117n^2 + 157n + 83)} = y_n. \quad (3.66)$$

Now, starting from (3.64) and taking into account (3.60)–(3.62), (3.66), and the assertions of Lemma 3.11, we obtain

$$|\widehat{W_i}| \le |\hat{z}_i - z_i| \left| \sum_{j=1}^{n} \frac{W_j}{\hat{z}_i - z_j} + 1 \right| \prod_{j \ne i} \left(1 + \frac{|\hat{z}_j - z_j|}{|\hat{z}_i - \hat{z}_j|} \right)$$
$$< \frac{\lambda_n}{c_n} |W_i| y_n \left(1 + \frac{\lambda_n}{1 - 2\lambda_n} \right)^{n-1}$$
$$= \phi_n |W_i|.$$

Using the symbolic computation in the programming package *Mathematica* 6.0, we find that the sequence $\{\phi_n\}_{n=3,4,\dots}$ attains its maximum for $n = 5$:

$$\phi_n \le \phi_5 < 0.22, \quad \text{for all } n \ge 3.$$

Therefore, $|\widehat{W_i}| < 0.22|W_i|$ and the assertion (i) is valid.

According to this, (3.63), and the inequality

$$\frac{0.22(3n+5)^3}{27n^3 + 99n^2 + 89n + 41} \le 0.32 < 1,$$

we find

$$|\widehat{W_i}| < 0.22|W_i| < 0.22 \frac{2d}{5n+3} < 0.22 \frac{2}{5n+3} \frac{(3n+5)^3}{27n^3 + 99n^2 + 89n + 41} \hat{d},$$

wherefrom

$$\hat{w} < \frac{2}{5n+3} \hat{d},$$

which proves the assertion (ii) of Lemma 3.12. □

Now, we are able to establish the main convergence theorem for the iterative method (3.55).

Theorem 3.8. *If the initial approximations $z_1^{(0)}, \dots, z_n^{(0)}$ satisfy the initial condition*

$$w^{(0)} < c_n d^{(0)}, \quad c_n = \frac{2}{5n+3} \quad (n \ge 3), \tag{3.67}$$

then the iterative method (3.55) is convergent.

Proof. It is sufficient to prove that the inequalities (i) and (ii) of Theorem 3.1 are valid for the correction

$$C_i^{(m)} = \frac{W_i^{(m)}}{1 + G_{1,i}^{(m)}} \left(1 - \frac{W_i^{(m)} G_{2,i}^{(m)}}{(1 + G_{1,i}^{(m)})^2} \right) \quad (i \in \boldsymbol{I}_n),$$

which appears in the considered method (3.55).

Using Lemma 3.11(i) and (3.59), we find

$$
\begin{aligned}
|C_i| = |\hat{z}_i - z_i| &= \left| \frac{W_i}{1 + G_{1,i}} \left(1 - \frac{W_i G_{2,i}}{(1 + G_{1,i})^2} \right) \right| \\
&< \frac{5n + 3}{3n + 5} W_i \left(1 + \frac{4(n - 1)}{(3n + 5)^2} \right) = \frac{(5n + 3)(9n^2 + 34n + 21)}{(3n + 5)^3} |W_i| \\
&= x_n |W_i|.
\end{aligned}
$$

It is easy to show that the sequence $\{x_n\}_{n=3,4,\ldots}$ is monotonically increasing and $x_n < \lim_{m \to \infty} x_n = 5/3$, wherefrom

$$
|C_i| < \frac{5}{3} |W_i|. \tag{3.68}
$$

In Lemma 3.12 (assertion (ii)), the implication $w < c_n d \Rightarrow \hat{w} < c_n \hat{d}$ has been proved. Using a similar procedure, we prove by induction that the initial condition (3.67) implies the inequality $w^{(m)} < c_n d^{(m)}$ for each $m = 1, 2, \ldots$. Therefore, by (i) of Lemma 3.12, we obtain

$$
|W_i^{(m+1)}| < 0.22 |W_i^{(m)}| \quad (i \in I_n, \ m = 0, 1, \ldots).
$$

According to this and by the inequalities (i) of Lemma 3.11 and (3.68), we obtain (omitting iteration indices)

$$
\begin{aligned}
|\widehat{C}_i| = \frac{5}{3} |\widehat{W}_i| &< \frac{5}{3} \cdot 0.22 |W_i| \\
&= \frac{1.1}{3} \left| \frac{W_i}{1 + G_{1,i}} \left(1 - \frac{W_i G_{2,i}}{(1 + G_{1,i})^2} \right) \right| \left| \frac{1 + G_{1,i}}{1 - \dfrac{W_i G_{2,i}}{(1 + G_{1,i})^2}} \right| \\
&< \frac{1.1}{3} |C_i| \frac{\dfrac{7n + 1}{5n + 3}}{1 - \dfrac{4(n - 1)}{(3n + 5)^2}} < 0.52 |C_i|.
\end{aligned}
$$

In this manner, we have proved by induction that the inequality $|C_i^{(m+1)}| < 0.52 |C_i^{(m)}|$ holds for each $i = 1, \ldots, n$ and $m = 0, 1, \ldots$. Furthermore, comparing this result with (i) of Theorem 3.1, we see that $\beta = 0.52 < 1$. This yields the first part of the theorem. In addition, according to (3.57), we note that the following is valid:

$$
|G_{1,i}| < (n - 1)c_n = \frac{2(n - 1)}{5n + 3} \le \frac{2}{9} < 1,
$$

which means that $0 \notin 1 + G_{1,i}$. Using induction and the assertion (ii) of Lemma 3.12, we prove that $0 \notin 1 + G_{1,i}^{(m)}$ for arbitrary iteration index m.

Therefore, under the condition (3.67), the iterative method (3.55) is well defined in each iteration.

To prove (ii) of Theorem 3.8, we first note that $\beta = 0.52$ yields $g(\beta) = \frac{1}{1-0.52} \approx 2.08$. It remains to prove the disjunctivity of the inclusion disks

$$S_1 = \{z_1^{(0)}; 2.08|C_1^{(0)}|\}, \ldots, S_n = \{z_n^{(0)}; 2.08|C_n^{(0)}|\}.$$

By virtue of (3.68), we have $|C_i^{(0)}| < \frac{5}{3}w^{(0)}$, wherefrom

$$d^{(0)} > \frac{5n+3}{2}w^{(0)} > \frac{5n+3}{2} \cdot \frac{3}{5}|C_i^{(0)}| \geq \frac{3(5n+3)}{20}\left(|C_i^{(0)}| + |C_j^{(0)}|\right)$$
$$> g(0.52)\left(|C_i^{(0)}| + |C_j^{(0)}|\right).$$

This means that

$$\left|z_i^{(0)} - z_j^{(0)}\right| \geq d^{(0)} > g(0.52)\left(|C_i^{(0)}| + |C_j^{(0)}|\right) = \operatorname{rad} S_i + \operatorname{rad} S_j.$$

Therefore, the inclusion disks $S_1, \ldots, S_n$ are disjoint, which completes the proof of Theorem 3.8. $\square$

3.3 Guaranteed Convergence: Sequence Approach

In what follows, we will present another concept of the convergence analysis involving initial conditions of the form (3.7) which guarantee the convergence of the considered methods.

Let $z_1^{(m)}, \ldots, z_n^{(m)}$ be approximations to the simple zeros $\zeta_1, \ldots, \zeta_n$ of a polynomial P, generated by some iterative method for the simultaneous determination of zeros at the mth iterative step and let $u_i^{(m)} = z_i^{(m)} - \zeta_i$ $(i \in \boldsymbol{I}_n)$. Our main aim is to study the convergence of the sequences $\{u_1^{(m)}\}, \ldots, \{u_n^{(m)}\}$ under the initial condition (3.7). In our analysis, we will use Corollary 1.1 proved in [118] (see Sect. 1.2).

The point estimation approach presented in this section consists of the following main steps:

1° If $c_n \leq 1/(2n)$ and (3.7) holds, from Corollary 1.1, it follows that the inequalities

$$|u_i^{(0)}| = |z_i^{(0)} - \zeta_i| < \frac{|W_i^{(0)}|}{1 - nc_n} \tag{3.69}$$

are valid for each $i \in \boldsymbol{I}_n$. These inequalities have an important role in the estimation procedure involved in the convergence analysis of the sequences $\{z_i^{(m)}\}$, produced by the considered simultaneous method.

2° In the next step, we derive the inequalities

$$d < \tau_n \hat{d} \quad \text{and} \quad |\widehat{W}_i| < \beta_n |W_i|,$$

which involve the minimal distances and the Weierstrass' corrections at two successive iterative steps. The i-factor c_n appearing in (3.7) has to be chosen to provide such values of τ_n and β_n which give the following implication

$$w < c_n d \implies \hat{w} < c_n \hat{d},$$

important in the proof of convergence theorems by induction. Let us note that the above implication will hold if $\tau_n \beta_n < 1$.

3° In the final step, we derive the inequalities of the form

$$|u_i^{(m+1)}| \leq \gamma(n, d^{(m)}) |u_i^{(m)}|^p \left(\sum_{\substack{j=1 \\ j \neq i}}^{n} |u_j^{(m)}|^q \right)^r \tag{3.70}$$

for $i = 1, \ldots, n$ and $m = 0, 1, \ldots,$ and prove that all sequences $\{|u_1^{(m)}|\}, \ldots, \{|u_n^{(m)}|\}$ tend to 0 under the condition (3.7) (with suitably chosen c_n), which means that $z_i^{(m)} \to \zeta_i$ ($i \in I_n$). The order of convergence of these sequences is obtained from (3.70) and it is equal to $p + qr$.

To study iterative methods which do not involve Weierstrass' corrections W_i, appearing in the initial conditions of the form (3.7), it is necessary to establish a suitable relation between Newton's correction $P(z_i)/P'(z_i)$ and Weierstrass' correction W_i. Applying the logarithmic derivative to $P(t)$, represented by the Lagrangean interpolation formula (3.27) (for $z = t$), one obtains

$$\frac{P'(t)}{P(t)} = \sum_{\substack{j=1 \\ j \neq i}}^{n} \frac{1}{t - z_j} + \frac{\displaystyle\sum_{\substack{j=1 \\ j \neq i}}^{n} \frac{W_j}{t - z_j} + 1 - (t - z_i) \sum_{\substack{j=1 \\ j \neq i}}^{n} \frac{W_j}{(t - z_j)^2}}{W_i + (t - z_i) \left[\displaystyle\sum_{\substack{j=1 \\ j \neq i}}^{n} \frac{W_j}{t - z_j} + 1 \right]}. \tag{3.71}$$

Putting $t = z_i$ in (3.71), we get Carstensen's identity [15]

$$\frac{P'(z_i)}{P(z_i)} = \sum_{\substack{j=1 \\ j \neq i}}^{n} \frac{1}{z_i - z_j} + \frac{\displaystyle\sum_{\substack{j=1 \\ j \neq i}}^{n} \frac{W_j}{z_i - z_j} + 1}{W_i}. \tag{3.72}$$

In what follows, we will apply the three-stage aforementioned procedure for the convergence analysis of some frequently used simultaneous methods.

The Ehrlich–Aberth's Method

In this part, we use Newton's and Weierstrass' correction, given, respectively, by

$$N_i^{(m)} = \frac{P(z_i^{(m)})}{P'(z_i^{(m)})} \quad \text{and} \quad W_i^{(m)} = \frac{P(z_i^{(m)})}{\prod\limits_{\substack{j=1 \\ j \neq i}}^{n} (z_i^{(m)} - z_j^{(m)})} \quad (i \in \boldsymbol{I}_n, \ m = 0, 1, \ldots).$$

We are concerned here with one of the most efficient numerical methods for the simultaneous approximation of all zeros of a polynomial, given by the iterative formula

$$z_i^{(m+1)} = z_i^{(m)} - \frac{1}{\dfrac{1}{N_i^{(m)}} - \sum\limits_{\substack{j=1 \\ j \neq i}}^{n} \dfrac{1}{z_i^{(m)} - z_j^{(m)}}} \quad (i \in \boldsymbol{I}_n, \ m = 0, 1, \ldots). \quad (3.73)$$

Our aim is to state practically applicable initial conditions of the form (3.7), which enable a guaranteed convergence of the Ehrlich–Aberth's method (3.73). First, we present a lemma concerned with the localization of polynomial zeros.

Lemma 3.13. *Assume that the following condition*

$$w < c_n d, \quad c_n = \begin{cases} \dfrac{1}{2n + 1.4}, & 3 \leq n \leq 7 \\[2mm] \dfrac{1}{2n}, & n \geq 8 \end{cases} \quad (3.74)$$

is satisfied. Then, each disk $\left\{ z_i; \dfrac{1}{1 - n c_n} |W_i| \right\}$ $(i \in \boldsymbol{I}_n)$ *contains one and only one zero of* P.

The above assertion follows from Corollary 1.1 under the condition (3.74).

Lemma 3.14. *Let* $z_1, \ldots, z_n$ *be distinct approximations to the zeros* $\zeta_1, \ldots, \zeta_n$ *of a polynomial* P *of degree* n *and let* $\hat{z}_1, \ldots, \hat{z}_n$ *be new respective approximations obtained by the Ehrlich–Aberth's method (3.73). Then, the following formula is valid:*

$$\widehat{W}_i = -(\hat{z}_i - z_i)^2 \sum_{j \neq i} \frac{W_j}{(\hat{z}_i - z_j)(z_i - z_j)} \prod_{j \neq i} \left(1 + \frac{\hat{z}_j - z_j}{\hat{z}_i - \hat{z}_j} \right). \quad (3.75)$$

Proof. From the iterative formula (3.73), one obtains

$$\frac{1}{\hat{z}_i - z_i} = \sum_{j \neq i} \frac{1}{z_i - z_j} - \frac{P'(z_i)}{P(z_i)},$$

so that, using (3.72),

$$\frac{W_i}{\hat{z}_i - z_i} = W_i \left(\sum_{j \neq i} \frac{1}{z_i - z_j} - \frac{P'(z_i)}{P(z_i)} \right) = -W_i \left[\frac{1}{W_i} \left(\sum_{j \neq i} \frac{W_j}{z_i - z_j} + 1 \right) \right]$$

$$= - \sum_{j \neq i} \frac{W_j}{z_i - z_j} - 1.$$

According to this, we have

$$\sum_{j=1}^{n} \frac{W_j}{\hat{z}_i - z_j} + 1 = \frac{W_i}{\hat{z}_i - z_i} + \sum_{j \neq i} \frac{W_j}{\hat{z}_i - z_j} + 1 = - \sum_{j \neq i} \frac{W_j}{z_i - z_j} + \sum_{j \neq i} \frac{W_j}{\hat{z}_i - z_j}$$

$$= -(\hat{z}_i - z_i) \sum_{j \neq i} \frac{W_j}{(\hat{z}_i - z_j)(z_i - z_j)}.$$

Taking into account the last expression, returning to (3.27), we find for $z = \hat{z}_i$

$$P(\hat{z}_i) = \left(\sum_{j=1}^{n} \frac{W_j}{\hat{z}_i - z_j} + 1 \right) \prod_{j=1}^{n} (\hat{z}_i - z_j)$$

$$= -(\hat{z}_i - z_i)^2 \sum_{j \neq i} \frac{W_j}{(\hat{z}_i - z_j)(z_i - z_j)} \prod_{j \neq i} (\hat{z}_i - z_j).$$

After dividing by $\prod_{j \neq i}(\hat{z}_i - \hat{z}_j)$ and some rearrangement, we obtain (3.75).
                                                                                    □

Let us introduce the abbreviations:

$$\rho_n = \frac{1}{1 - n c_n}, \quad \gamma_n = \frac{1}{1 - \rho_n c_n - (n-1)(\rho_n c_n)^2},$$

$$\lambda_n = \rho_n c_n (1 - \rho_n c_n) \gamma_n, \quad \beta_n = \frac{(n-1)\lambda_n^2}{1 - \lambda_n} \left(1 + \frac{\lambda_n}{1 - 2\lambda_n} \right)^{n-1}.$$

Lemma 3.15. *Let $\hat{z}_1, \ldots, \hat{z}_n$ be approximations produced by the Ehrlich–Aberth's method (3.73) and let $u_i = z_i - \zeta_i$ and $\hat{u}_i = \hat{z}_i - \zeta_i$. If $n \geq 3$ and the inequality (3.74) holds, then:*

(i) $d < \dfrac{1}{1 - 2\lambda_n} \hat{d}.$

(ii) $\widehat{w} < \beta_n w.$

(iii) $\widehat{w} < c_n \widehat{d}.$

(iv) $|\widehat{u}_i| \leq \frac{\gamma_n}{d^2}|u_i|^2 \sum_{j \neq i} |u_j|.$

Proof. According to the initial condition (3.74) and Lemma 3.13, we have

$$|u_i| = |z_i - \zeta_i| \leq \rho_n |W_i| \leq \rho_n w < \rho_n c_n d. \tag{3.76}$$

In view of (3.76) and the definition of the minimal distance d, we find

$$|z_j - \zeta_i| \geq |z_j - z_i| - |z_i - \zeta_i| > d - \rho_n c_n d = (1 - \rho_n c_n)d. \tag{3.77}$$

Using the identity

$$\frac{P'(z_i)}{P(z_i)} = \sum_{j=1}^{n} \frac{1}{z_i - \zeta_j} = \frac{1}{u_i} + \sum_{j \neq i} \frac{1}{z_i - \zeta_j}, \tag{3.78}$$

from (3.73), we get

$$\widehat{u}_i = \widehat{z}_i - \zeta_i = z_i - \zeta_i - \cfrac{1}{\cfrac{1}{u_i} + \sum_{j \neq i} \cfrac{1}{z_i - \zeta_j} - \sum_{j \neq i} \cfrac{1}{z_i - z_j}}$$

$$= u_i - \frac{u_i}{1 - u_i S_i} = -\frac{u_i^2 S_i}{1 - u_i S_i}, \tag{3.79}$$

where $S_i = \sum_{j \neq i} \dfrac{u_j}{(z_i - \zeta_j)(z_i - z_j)}.$

Using the definition of d and the bounds (3.76) and (3.77), we estimate

$$|u_i S_i| \leq |u_i| \sum_{j \neq i} \frac{|u_j|}{|z_i - \zeta_j||z_i - z_j|} < \rho_n c_n d \cdot \frac{(n-1)\rho_n c_n d}{(1 - \rho_n c_n)d^2}$$

$$= \frac{(\rho_n c_n)^2 (n-1)}{1 - \rho_n c_n}. \tag{3.80}$$

Now, by (3.76) and (3.80), from (3.73) we find

$$|\widehat{z}_i - z_i| = \left| \frac{u_i}{1 - u_i S_i} \right| \leq \frac{|u_i|}{1 - |u_i S_i|} < \frac{|u_i|}{1 - \dfrac{(\rho_n c_n)^2 (n-1)}{1 - \rho_n c_n}}$$

$$< \frac{\rho_n c_n (1 - \rho_n c_n)}{1 - \rho_n c_n - (\rho_n c_n)^2 (n-1)} d = \rho_n c_n (1 - \rho_n c_n) \gamma_n d$$

$$= \lambda_n d \tag{3.81}$$

and also
$$|\hat{z}_i - z_i| < (1 - \rho_n c_n)\gamma_n |u_i| < (1 - \rho_n c_n)\gamma_n \rho_n |W_i|. \tag{3.82}$$

Having in mind (3.81), according to Lemma 3.2, we conclude that the estimates $|\hat{z}_i - z_j| > (1 - \lambda_n)d$ and $|\hat{z}_i - \hat{z}_j| > (1 - 2\lambda_n)d$ ($i \in \boldsymbol{I}_n$) hold. From the last inequality, we find
$$\frac{d}{\hat{d}} < \frac{1}{1 - 2\lambda_n} \quad \text{for every } n \geq 3, \tag{3.83}$$

which proves the assertion (i) of Lemma 3.15.

Using the starting inequality $w/d < c_n$ and the bounds (3.81), (3.82), (3.14), (3.15), and (3.16), we estimate the quantities involved in (3.75):

$$\begin{aligned}
|\widehat{W}_i| &\leq |\hat{z}_i - z_i|^2 \sum_{j \neq i} \frac{|W_j|}{|\hat{z}_i - z_j||z_i - z_j|} \prod_{j \neq i}\left(1 + \frac{|\hat{z}_j - z_j|}{|\hat{z}_i - \hat{z}_j|}\right) \\
&< \frac{(n-1)\lambda_n^2}{(1 - \lambda_n)}\left(1 + \frac{\lambda_n}{1 - 2\lambda_n}\right)^{n-1} |W_i| \\
&= \beta_n |W_i|.
\end{aligned}$$

Therefore, we have
$$\widehat{w} < \beta_n w \tag{3.84}$$

so that, by (3.74), (3.83), and (3.84), we estimate
$$\widehat{w} < \beta_n w < \beta_n c_n d < \frac{\beta_n}{1 - 2\lambda_n} c_n \hat{d}.$$

Since
$$\frac{\beta_n}{1 - 2\lambda_n} < 0.95 < 1 \quad \text{for all} \quad 3 \leq n \leq 7$$

and
$$\frac{\beta_n}{1 - 2\lambda_n} < 0.78 < 1 \quad \text{for all} \quad n \geq 8,$$

we have
$$\widehat{w} < c_n \hat{d}, \quad n \geq 3.$$

In this way, we have proved the assertions (ii) and (iii) of Lemma 3.15.

Using the previously derived bounds, we find
$$\begin{aligned}
|\hat{u}_i| &\leq \frac{|u_i|^2 |S_i|}{1 - |u_i S_i|} < \frac{|u_i|^2}{1 - \dfrac{(\rho c_n)^2 (n-1)}{1 - \rho_n c_n}} \sum_{j \neq i} \frac{|u_j|}{|z_i - \zeta_j||z_i - z_j|} \\
&< \frac{1 - \rho_n c_n}{1 - \rho_n c_n - (\rho_n c_n)^2 (n-1)} |u_i|^2 \sum_{j \neq i} \frac{|u_j|}{(1 - \rho_n c_n)d^2} \\
&= \frac{1}{(1 - \rho_n c_n - (\rho_n c_n)^2 (n-1))d^2} |u_i|^2 \sum_{j \neq i} |u_j|,
\end{aligned}$$

wherefrom

$$|\hat{u}_i| < \frac{\gamma_n}{d^2}|u_i|^2 \sum_{j \neq i} |u_j|. \tag{3.85}$$

This strict inequality is derived assuming that $u_i \neq 0$ (see Remark 3.8). If we include the case $u_i = 0$, then it follows

$$|\hat{u}_i| \leq \frac{\gamma_n}{d^2}|u_i|^2 \sum_{j \neq i} |u_j|$$

and the assertion (iv) of Lemma 3.15 is proved. $\square$

Remark 3.7. In what follows, the assertions of the form (i)–(iv) of Lemma 3.15 will be presented for the three other methods, but for different i-factor c_n and specific entries of λ_n, β_n, and γ_n.

We now give the convergence theorem for the Ehrlich–Aberth's method (3.73), which involves only initial approximations to the zeros, the polynomial coefficients, and the polynomial degree n.

Theorem 3.9. *Under the initial condition*

$$w^{(0)} < c_n d^{(0)}, \tag{3.86}$$

where c_n is given by (3.74), the Ehrlich–Aberth's method (3.73) is convergent with the cubic convergence.

Proof. The convergence analysis is based on the estimation procedure of the errors $u_i^{(m)} = z_i^{(m)} - \zeta_i$ ($i \in I_n$). The proof is by induction with the argumentation used for the inequalities (i)–(iv) of Lemma 3.15. Since the initial condition (3.86) coincides with (3.74), all estimates given in Lemma 3.15 are valid for the index $m = 1$. Actually, this is the part of the proof with respect to $m = 1$. Furthermore, the inequality (iii) again reduces to the condition of the form (3.74) and, therefore, the assertions (i)–(iv) of Lemma 3.15 hold for the next index, and so on. All estimates and bounds for the index m are derived essentially in the same way as for $m = 0$. In fact, the implication

$$w^{(m)} < c_n d^{(m)} \implies w^{(m+1)} < c_n d^{(m+1)}$$

plays the key role in the convergence analysis of the Ehrlich–Aberth's method (3.73) because it involves the initial condition (3.86), which enables the validity of all inequalities given in Lemma 3.15 for all $m = 0, 1, \ldots$. In particular, regarding (3.83) and (3.85), we have

$$\frac{d^{(m)}}{d^{(m+1)}} < \frac{1}{1 - 2\lambda_n} \tag{3.87}$$

and

$$|u_i^{(m+1)}| \leq \frac{\gamma_n}{(d^{(m)})^2}|u_i^{(m)}|^2 \sum_{\substack{j=1 \\ j \neq i}}^{n}|u_j^{(m)}| \quad (i \in \boldsymbol{I}_n) \tag{3.88}$$

for each iteration index $m = 0, 1, \ldots$ if (3.86) holds.

Substituting

$$t_i^{(m)} = \left[\frac{(n-1)\gamma_n}{(1-2\lambda_n)(d^{(m)})^2}\right]^{1/2}|u_i^{(m)}|,$$

the inequalities (3.88) become

$$t_i^{(m+1)} \leq \frac{(1-2\lambda_n)d^{(m)}}{(n-1)d^{(m+1)}}\left(t_i^{(m)}\right)^2 \sum_{\substack{j=1 \\ j \neq i}}^{n} t_j^{(m)},$$

wherefrom, by (3.87),

$$t_i^{(m+1)} < \frac{\left(t_i^{(m)}\right)^2}{n-1} \sum_{\substack{j=1 \\ j \neq i}}^{n} t_j^{(m)} \quad (i \in \boldsymbol{I}_n). \tag{3.89}$$

By virtue of (3.76), we find

$$t_i^{(0)} = \sqrt{\frac{(n-1)\gamma_n}{(1-2\lambda_n)(d^{(0)})^2}}|u_i^{(0)}| < \rho_n c_n d^{(0)}\sqrt{\frac{(n-1)\gamma_n}{(1-2\lambda_n)(d^{(0)})^2}}$$

$$= \rho_n c_n \sqrt{\frac{(n-1)\gamma_n}{1-2\lambda_n}}$$

for each $i = 1, \ldots, n$. Taking

$$t = \max_{1 \leq i \leq n} t_i^{(0)} < \rho_n c_n \sqrt{\frac{(n-1)\gamma_n}{1-2\lambda_n}},$$

we come to the inequalities

$$t_i^{(0)} \leq t < 0.571 < 1 \quad (3 \leq n \leq 7)$$

and

$$t_i^{(0)} \leq t < 0.432 < 1 \quad (n \geq 8)$$

for all $i = 1, \ldots, n$. According to this, from (3.89), we conclude that the sequences $\{t_i^{(m)}\}$ (and, consequently, $\{|u_i^{(m)}|\}$) tend to 0 for all $i = 1, \ldots, n$. Therefore, the Ehrlich–Aberth's method (3.73) is convergent.

Taking into account that the quantity $d^{(m)}$, which appears in (3.88), is bounded (see the proof of Theorem 5.1) and tends to $\min_{i \neq j}|\zeta_i - \zeta_j|$ and setting

$$u^{(m)} = \max_{1 \le i \le n} |u_i^{(m)}|,$$

from (3.88), we obtain

$$|u_i^{(m+1)}| \le u^{(m+1)} < \frac{(n-1)\gamma_n}{d^{(m)}} |u^{(m)}|^3,$$

which proves the cubic convergence.   □

Remark 3.8. As usual in the convergence analysis of iterative methods (see, e.g., [48]), we could assume that the errors $u_i^{(m)} = z_i^{(m)} - \zeta_i$ ($i \in \boldsymbol{I}_n$) do not reach 0 for a finite m. However, if $u_i^{(m_0)} = 0$ for some indices $i_1, \dots, i_k$ and $m_0 \ge 0$, we just take $z_{i_1}^{(m_0)}, \dots, z_{i_k}^{(m_0)}$ as approximations to the zeros $\zeta_{i_1}, \dots, \zeta_{i_k}$ and do not iterate further for the indices $i_1, \dots, i_k$. If the sequences $\{u_i^{(m)}\}$ ($i \in \boldsymbol{I}_n \setminus \{i_1, \dots, i_k\}$) have the order of convergence q, then obviously the sequences $\{u_{i_1}^{(m)}\}, \dots, \{u_{i_k}^{(m)}\}$ converge with the convergence rate at least q. This remark refers not only to the iterative method (3.73) but also to all methods considered in this book. For this reason, we do not discuss this point further.

The Ehrlich–Aberth's Method with Newton's Corrections

The convergence of the Ehrlich–Aberth's method (3.1) can be accelerated using Newton's corrections $N_i^{(m)} = P(z_i^{(m)})/P'(z_i^{(m)})$ ($i \in \boldsymbol{I}_n$, $m = 0, 1, \dots$). In this way, the following method for the simultaneous approximation of all simple zeros of a given polynomial P can be established

$$z_i^{(m+1)} = z_i^{(m)} - \frac{1}{\dfrac{1}{N_i^{(m)}} - \sum_{j \ne i} \dfrac{1}{z_i^{(m)} - z_j^{(m)} + N_j^{(m)}}} \qquad (i \in \boldsymbol{I}_n), \qquad (3.90)$$

where $m = 0, 1, \dots$, see Sect. 1.1. This method will be briefly called the EAN method.

From Corollary 1.1, the following lemma can be stated.

Lemma 3.16. *Let* $z_1, \dots, z_n$ *be distinct numbers satisfying the inequality*

$$w < c_n d, \quad c_n = \begin{cases} \dfrac{1}{2.2n + 1.9}, & 3 \le n \le 21 \\[2mm] \dfrac{1}{2.2n}, & n \ge 22 \end{cases} . \qquad (3.91)$$

Then, the disks $\left\{ z_1; \dfrac{1}{1 - nc_n} |W_1| \right\}, \dots, \left\{ z_n; \dfrac{1}{1 - nc_n} |W_n| \right\}$ *are mutually disjoint and each of them contains exactly one zero of a polynomial P.*

We now give the expression for the improved Weierstrass' correction $\widehat{W}_i$.

Lemma 3.17. *Let $z_1, \ldots, z_n$ be distinct approximations to the zeros $\zeta_1, \ldots, \zeta_n$ of a polynomial P of degree n and let $\hat{z}_1, \ldots, \hat{z}_n$ be new respective approximations obtained by the EAN method (3.90). Then, the following formula is valid:*

$$\widehat{W}_i = -(\hat{z}_i - z_i)\Big(W_i \Sigma_{N,i} + (\hat{z}_i - z_i)\Sigma_{W,i}\Big)\prod_{j\neq i}\Big(1 + \frac{\hat{z}_j - z_j}{\hat{z}_i - \hat{z}_j}\Big), \qquad (3.92)$$

where

$$\Sigma_{N,i} = \sum_{j\neq i} \frac{N_j}{(z_i - z_j + N_j)(z_i - z_j)}, \quad \Sigma_{W,i} = \sum_{j\neq i} \frac{W_j}{(\hat{z}_i - z_j)(z_i - z_j)}.$$

The relation (3.92) is obtained by combining the Lagrangean interpolation formula (3.27) for $z = \hat{z}_i$, the iterative formula (3.90), and the identity (3.72). Since the proving technique of Lemma 3.17 is a variation on earlier procedure applied in the proof of Lemma 3.14, we shall pass over it lightly. The complete proof can be found in [119].

We introduce the abbreviations:

$$\rho_n = \frac{1}{1 - nc_n}, \quad \delta_n = 1 - \rho_n c_n - (n-1)\rho_n c_n,$$

$$\alpha_n = (1 - \rho_n c_n)((1 - \rho_n c_n)^2 - (n-1)\rho_n c_n),$$

$$\gamma_n = \frac{n-1}{\alpha_n - (n-1)^2(\rho_n c_n)^3}, \quad \lambda_n = \frac{\alpha_n \gamma_n \rho_n c_n}{n-1},$$

$$\beta_n = \lambda_n(n-1)\Big(\frac{(1 - \rho_n c_n)^2 \rho_n c_n}{\alpha_n} + \frac{\lambda_n}{1 - \lambda_n}\Big)\Big(1 + \frac{\lambda_n}{1 - 2\lambda_n}\Big)^{n-1}.$$

Lemma 3.18. *Let $\hat{z}_1, \ldots, \hat{z}_n$ be approximations generated by the EAN method (3.90) and let $u_i = z_i - \zeta_i$, $\hat{u}_i = \hat{z}_i - \zeta_i$. If $n \geq 3$ and the inequality (3.91) holds, then:*

(i) $d < \dfrac{1}{1 - 2\lambda_n}\hat{d}.$

(ii) $\widehat{w} < \beta_n w.$

(iii) $\widehat{w} < c_n \hat{d}.$

(iv) $|\hat{u}_i| \leq \dfrac{\gamma_n}{d^3}|u_i|^2 \sum_{j\neq i}|u_j|^2.$

Proof. In regard to (3.91) and Lemma 3.16, we have $\zeta_i \in \Big\{z_i; \dfrac{1}{1 - nc_n}|W_i|\Big\}$ $(i \in \mathbf{I}_n)$, so that

$$|u_i| = |z_i - \zeta_i| \leq \rho_n|W_i| \leq \rho_n w < \rho_n c_n d. \qquad (3.93)$$

According to this and the definition of the minimal distance d, we find

$$|z_j - \zeta_i| \geq |z_j - z_i| - |z_i - \zeta_i| > d - \rho_n c_n d = (1 - \rho_n c_n) d. \tag{3.94}$$

Using the identity (3.78) and the estimates (3.93) and (3.94), we obtain

$$\left| \frac{P'(z_i)}{P(z_i)} \right| = \left| \sum_{j=1}^{n} \frac{1}{z_i - \zeta_j} \right| \geq \frac{1}{|z_i - \zeta_i|} - \sum_{j \neq i} \frac{1}{|z_i - \zeta_j|} > \frac{1}{\rho_n c_n d} - \frac{n-1}{(1 - \rho_n c_n) d}$$

$$= \frac{1 - \rho_n c_n - (n-1)\rho_n c_n}{(1 - \rho_n c_n) \rho_n c_n d} = \frac{\delta_n}{(1 - \rho_n c_n) \rho_n c_n d}.$$

Hence

$$|N_i| = \left| \frac{P(z_i)}{P'(z_i)} \right| < \frac{(1 - \rho_n c_n) \rho_n c_n d}{\delta_n}, \tag{3.95}$$

so that

$$|z_i - z_j + N_j| \geq |z_i - z_j| - |N_j| > d - \frac{(1 - \rho_n c_n) \rho_n c_n d}{\delta_n}$$

$$= \frac{(1 - \rho_n c_n)^2 - (n-1)\rho_n c_n}{\delta_n} d = \frac{\alpha_n}{\delta_n (1 - \rho_n c_n)} d. \tag{3.96}$$

Let us introduce

$$S_i = \sum_{j \neq i} \frac{N_j - u_j}{(z_i - \zeta_j)(z_i - z_j + N_j)}, \quad h_j = \sum_{k \neq j} \frac{1}{z_j - \zeta_k}.$$

We start from the iterative formula (3.90) and use the identity (3.78) to find

$$\hat{u}_i = \hat{z}_i - \zeta_i = z_i - \zeta_i - \frac{1}{\dfrac{1}{u_i} + \sum_{j \neq i} \dfrac{1}{z_i - \zeta_j} - \sum_{j \neq i} \dfrac{1}{z_i - z_j + N_j}}$$

$$= u_i - \frac{u_i}{1 + u_i \sum_{j \neq i} \dfrac{N_j - u_j}{(z_i - \zeta_j)(z_i - z_j + N_j)}}$$

$$= u_i - \frac{u_i}{1 + u_i S_i} = \frac{u_i^2 S_i}{1 + u_i S_i}. \tag{3.97}$$

Furthermore, we find

$$N_j = \frac{u_j}{1 + u_j h_j}, \quad N_j - u_j = -\frac{u_j^2 h_j}{1 + u_j h_j}, \quad S_i = -\sum_{j \neq i} \frac{\dfrac{u_j^2 h_j}{1 + u_j h_j}}{(z_i - \zeta_j)(z_i - z_j + N_j)}.$$

Using (3.93) and the inequality

$$|h_j| = \left| \sum_{k \neq j} \frac{1}{z_j - \zeta_k} \right| < \frac{n-1}{(1 - \rho_n c_n)d},$$

we find

$$\left| \frac{h_j}{1 + u_j h_j} \right| \leq \frac{|h_j|}{1 - |u_j||h_j|} < \frac{\dfrac{n-1}{(1 - \rho_n c_n)d}}{1 - \rho_n c_n d \dfrac{n-1}{(1 - \rho_n c_n)d}} = \frac{n-1}{\delta_n d}. \tag{3.98}$$

Combining (3.93), (3.94), (3.96), and (3.98), we obtain

$$|u_i S_i| \leq |u_i| \sum_{j \neq i} \frac{|u_j|^2 \left| \dfrac{h_j}{1 + u_j h_j} \right|}{|z_i - \zeta_j||z_i - z_j + N_j|}$$

$$< \rho_n c_n d \cdot \frac{(n-1)(\rho_n c_n d)^2 \dfrac{n-1}{\delta_n d}}{(1 - \rho_n c_n)d \dfrac{\alpha_n}{\delta_n(1 - \rho_n c_n)} d}$$

$$= \frac{(n-1)^2 (\rho_n c_n)^3}{\alpha_n}. \tag{3.99}$$

Using (3.93) and (3.99), from (3.90), we find

$$|\hat{z}_i - z_i| = \left| \frac{u_i}{1 + u_i S_i} \right| \leq \frac{|u_i|}{1 - |u_i S_i|} < \frac{|u_i|}{1 - \dfrac{(n-1)^2(\rho_n c_n)^3}{\alpha_n}}$$

$$= \frac{\alpha_n}{\alpha_n - (n-1)^2(\rho_n c_n)^3} |u_i| < \frac{\alpha_n \rho_n c_n \gamma_n}{n-1} d = \lambda_n d$$

and

$$|\hat{z}_i - z_i| < \frac{\alpha_n}{\alpha_n - (n-1)^2(\rho_n c_n)^3} |u_i| < \frac{\alpha_n \rho_n \gamma_n}{n-1} |W_i| = \frac{\lambda_n}{c_n} |W_i| < \lambda_n d. \tag{3.100}$$

Since (3.100) holds, we apply Lemma 3.2 and obtain

$$d < \frac{1}{1 - 2\lambda_n} \hat{d} \tag{3.101}$$

from (3.15). Thus, the assertion (i) of Lemma 3.18 is valid.

Using the starting inequality $w/d < c_n$ and the bounds (3.95), (3.96), (3.100), (3.14), and (3.15), for $n \geq 3$, we estimate the quantities appearing in (3.92):

$$|W_i||\Sigma_{N,i}| < w \frac{(n-1)\dfrac{(1-\rho_n c_n)\rho_n c_n d}{\delta_n}}{\dfrac{\alpha_n}{\delta_n(1-\rho_n c_n)}d^2} < \frac{(n-1)(1-\rho_n c_n)^2 \rho_n c_n^2}{\alpha_n},$$

$$|\hat{z}_i - z_i||\Sigma_{W,i}| < \lambda_n d \frac{(n-1)c_n d}{(1-\lambda_n)d \cdot d} < \frac{(n-1)\lambda_n c_n}{1-\lambda_n}.$$

According to the last two bounds and (3.16), from (3.92), we estimate

$$|\widehat{W}_i| \le |\hat{z}_i - z_i|\Big(|W_i||\Sigma_{N,i}| + |\hat{z}_i - z_i||\Sigma_{W,i}|\Big)\Big|\prod_{j \ne i}\Big(1 + \frac{\hat{z}_j - z_j}{\hat{z}_i - \hat{z}_j}\Big)\Big|$$

$$< \frac{\lambda_n}{c_n}|W_i|\Big(\frac{(n-1)(1-\rho_n c_n)^2 \rho_n c_n^2}{\alpha_n} + \frac{(n-1)\lambda_n c_n}{1-\lambda_n}\Big)\Big(1 + \frac{\lambda_n}{1+2\lambda_n}\Big)^{n-1}$$

$$= \beta_n|W_i| \le \beta_n w,$$

i.e.,

$$\widehat{w} < \beta_n w. \tag{3.102}$$

Therefore, we have proved the assertion (ii) of Lemma 3.18.

Since

$$\frac{\beta_n}{1-2\lambda_n} < 0.942 \qquad \text{for} \quad 3 \le n \le 21$$

and

$$\frac{\beta_n}{1-2\lambda_n} < 0.943 \qquad \text{for} \quad n \ge 22,$$

starting from (3.102), by (3.91) and (3.101), we find

$$\widehat{w} < \beta_n w < \beta_n c_n d < \frac{\beta_n}{1-2\lambda_n} \cdot c_n \hat{d} < c_n \hat{d},$$

which means that the implication $w < c_n d \Rightarrow \widehat{w} < c_n d$ holds. This proves (iii) of Lemma 3.18.

Using the above bounds, from (3.97), we obtain

$$|\hat{u}_i| \le \frac{|u_i|^2|S_i|}{1 - |u_i S_i|} < \frac{\alpha_n}{\alpha_n - (n-1)^2(\rho_n c_n)^3}|u_i|^2 \sum_{j \ne i} \frac{|u_j|^2\Big|\dfrac{h_j}{1+u_j h_j}\Big|}{|z_i - \zeta_j||z_i - z_j + N_j|}$$

$$< \frac{\alpha_n}{\alpha_n - (n-1)^2(\rho_n c_n)^3} \frac{\dfrac{n-1}{\delta_n d}}{(1-\rho_n c_n)d \dfrac{\alpha_n}{\delta_n(1-\rho_n c_n)}d}|u_i|^2 \sum_{j \ne i}|u_j|^2,$$

wherefrom (taking into account Remark 3.8)

$$|\hat{u}_i| \le \frac{\gamma_n}{d^3} |u_i|^2 \sum_{j \ne i} |u_j|^2,$$

which proves (iv) of Lemma 3.18.   □

Now, we give the convergence theorem for the EAN method (3.90).

Theorem 3.10. *Let P be a polynomial of degree $n \ge 3$ with simple zeros. If the initial condition*

$$w^{(0)} < c_n d^{(0)} \tag{3.103}$$

holds, where c_n is given by (3.91), then the EAN method (3.90) is convergent with the order of convergence 4.

Proof. Similarly to the proof of Theorem 3.9, we apply induction with the argumentation used for the inequalities (i)–(iv) of Lemma 3.18. According to (3.103) and (3.91), all estimates given in Lemma 3.18 are valid for the index $m = 1$. We notice that the inequality (iii) coincides with the condition of the form (3.103), and hence, the assertions (i)–(iv) of Lemma 3.18 are valid for the next index, etc. The implication

$$w^{(m)} < c_n d^{(m)} \implies w^{(m+1)} < c_n d^{(m+1)}$$

provides the validity of all inequalities given in Lemma 3.18 for all $m = 0, 1, \dots$. In particular, we have

$$\frac{d^{(m)}}{d^{(m+1)}} < \frac{1}{1 - 2\lambda_n} \tag{3.104}$$

and

$$|u_i^{(m+1)}| \le \frac{\gamma_n}{(d^{(m)})^3} |u_i^{(m)}|^2 \sum_{\substack{j=1 \\ j \ne i}}^{n} |u_j^{(m)}|^2 \quad (i \in \boldsymbol{I}_n) \tag{3.105}$$

for each iteration index $m = 0, 1, \dots$, where

$$\gamma_n = \frac{n-1}{\alpha_n - (n-1)^2 (\rho_n c_n)^3}.$$

Substituting

$$t_i^{(m)} = \left[\frac{(n-1)\gamma_n}{(1 - 2\lambda_n)(d^{(m)})^3} \right]^{1/3} |u_i^{(m)}|$$

into (3.105) yields

$$t_i^{(m+1)} \le \frac{(1 - 2\lambda_n)d^{(m)}}{(n-1)d^{(m+1)}} \left(t_i^{(m)}\right)^2 \sum_{\substack{j=1 \\ j \ne i}}^{n} \left(t_j^{(m)}\right)^2 \quad (i \in \boldsymbol{I}_n).$$

Hence, using (3.104), we obtain

$$t_i^{(m+1)} < \frac{1}{n-1}[t_i^{(m)}]^2 \sum_{\substack{j=1 \\ j \neq i}}^{n}[t_j^{(m)}]^2 \quad (i \in I_n).$$ (3.106)

Using (3.93), we find

$$t_i^{(0)} = \left[\frac{(n-1)\gamma_n}{(1-2\lambda_n)(d^{(0)})^3}\right]^{1/3}|u_i^{(0)}| < \rho_n c_n d^{(0)}\left[\frac{(n-1)\gamma_n}{(1-2\lambda_n)(d^{(0)})^3}\right]^{1/3}$$

$$= \rho_n c_n \left[\frac{(n-1)\gamma_n}{1-2\lambda_n}\right]^{1/3}.$$

Taking $t = \max\limits_{1 \leq i \leq n} t_i^{(0)}$ yields

$$t_i^{(0)} \leq t < 0.626 < 1 \quad (3 \leq n \leq 21)$$

and

$$t_i^{(0)} \leq t < 0.640 < 1 \quad (n \geq 22)$$

for each $i = 1, \ldots, n$. In regard to this, we conclude from (3.106) that the sequences $\{t_i^{(m)}\}$ and $\{|u_i^{(m)}|\}$ tend to 0 for all $i = 1, \ldots, n$, meaning that $z_i^{(m)} \to \zeta_i$. Therefore, the EAN method (3.90) is convergent. Besides, taking into account that the quantity $d^{(m)}$ appearing in (3.105) is bounded and tends to $\min\limits_{\substack{1 \leq i,j \leq n \\ i \neq j}} |\zeta_i - \zeta_j|$ and setting

$$u^{(m)} = \max\limits_{1 \leq i \leq n} |u_i^{(m)}|,$$

from (3.105), we obtain

$$|u_i^{(m+1)}| \leq u^{(m+1)} < (n-1)\frac{\gamma_n}{(d^{(m)})^3}(u^{(m)})^4,$$

which means that the order of convergence of the EAN method is 4. $\square$

The Börsch-Supan's Method with Weierstrass' Correction

The cubically convergent Börsch-Supan's method

$$z_i^{(m+1)} = z_i^{(m)} - \frac{W_i^{(m)}}{1 + \sum\limits_{j \neq i}\dfrac{W_j^{(m)}}{z_i^{(m)} - z_j^{(m)}}} \quad (i \in I_n, \, m = 0, 1, \ldots),$$

presented in [10], can be accelerated by using Weierstrass' corrections $W_i^{(m)} = P(z_i^{(m)})/\prod_{j\neq i}(z_i^{(m)} - z_j^{(m)})$. In this manner, we obtain the following iterative formula (see Nourein [95])

$$
z_i^{(m+1)} = z_i^{(m)} - \frac{W_i^{(m)}}{1 + \sum_{j\neq i} \dfrac{W_j^{(m)}}{z_i^{(m)} - W_i^{(m)} - z_j^{(m)}}} \qquad (i \in \boldsymbol{I}_n,\ m = 0, 1, \ldots).
$$

$$(3.107)$$

The order of convergence of the Börsch-Supan's method with Weierstrass' corrections (3.107) is 4 (see, e.g., [16], [188]). For brevity, the method (3.107) will be referred to as the BSW method.

Let us introduce the abbreviations:

$$
\rho_n = \frac{1}{1 - nc_n}, \quad \gamma_n = \frac{\rho_n(1 + \rho_n c_n)^{2n-2}}{(1 - \rho_n c_n)^2},
$$

$$
\lambda_n = \rho_n c_n (1 - c_n), \quad \beta_n = \frac{\lambda_n \rho_n c_n^2 (n-1)^2}{(1 - \lambda_n)(1 - c_n)}\left(1 + \frac{\lambda_n}{1 - 2\lambda_n}\right)^{n-1}.
$$

Lemma 3.19. *Let $\hat{z}_1, \ldots, \hat{z}_n$ be approximations obtained by the iterative method (3.107) and let $\hat{u}_i = \hat{z}_i - \zeta_i$, $\hat{d} = \min_{i\neq j} |\hat{z}_i - \hat{z}_j|$, and $\hat{w} = \max_{1\leq i \leq n} |\widehat{W}_i|$.*
If the inequality

$$
w < c_n d, \qquad c_n = \begin{cases} \dfrac{1}{2n+1}, & 3 \leq n \leq 13 \\[2mm] \dfrac{1}{2n}, & n \geq 14 \end{cases}
$$

$$(3.108)$$

holds, then:

(i) $\hat{w} < \beta_n w$.

(ii) $d < \dfrac{1}{1 - 2\lambda_n}\hat{d}$.

(iii) $|u_i| < \rho_n c_n d$.

(iv) $\hat{w} < c_n \hat{d}$.

(v) $|\hat{u}_i| \leq \dfrac{\gamma_n}{d^3}|u_i|^2\left(\sum_{j\neq i} |u_j|\right)^2$.

The proof of this lemma is strikingly similar to that of Lemmas 3.15 and 3.18 and will be omitted.

Now, we establish initial conditions of practical interest, which guarantee the convergence of the BSW method (3.107).

Theorem 3.11. *If the initial condition given by*

$$
w^{(0)} < c_n d^{(0)}
$$

$$(3.109)$$

is satisfied, where c_n is given by (3.108), then the iterative method (3.107) is convergent with the order of convergence 4.

Proof. The proof of this theorem is based on the assertions of Lemma 3.19 with the help of the previously presented technique. As in the already stated convergence theorems, the proof goes by induction. By the same argumentation as in the previous proofs, the initial condition (3.109) provides the validity of the inequality $w^{(m)} < c_n d^{(m)}$ for all $m \geq 0$, and hence, the inequalities (i)–(iv) of Lemma 3.19 also hold for all $m \geq 0$. In particular (according to Lemma 3.19(i)), we have

$$\frac{d^{(m)}}{d^{(m+1)}} < \frac{1}{1 - 2\lambda_n} \tag{3.110}$$

and, with regard to Lemma 3.19(iv),

$$|u_i^{(m+1)}| \leq \frac{\gamma_n}{(d^{(m)})^3} |u_i^{(m)}|^2 \left(\sum_{\substack{j=1 \\ j \neq i}}^{n} |u_j^{(m)}| \right)^2 \tag{3.111}$$

for each $i \in I_n$ and all $m = 0, 1, \ldots.$.
    Substituting

$$t_i^{(m)} = \left[\frac{(n-1)^2 \gamma_n}{(1 - 2\lambda_n)(d^{(m)})^3} \right]^{1/3} |u_i^{(m)}|$$

into (3.111) and using (3.110), we obtain

$$t_i^{(m+1)} < \frac{1}{(n-1)^2} (t_i^{(m)})^2 \left(\sum_{\substack{j=1 \\ j \neq i}}^{n} t_j^{(m)} \right)^2. \tag{3.112}$$

By the assertion (ii) of Lemma 3.19 for the first iteration ($m = 0$), we have

$$t_i^{(0)} = \left[\frac{(n-1)^2 \gamma_n}{(1 - 2\lambda_n)(d^{(0)})^3} \right]^{1/3} |u_i^{(0)}| < \rho_n c_n \left[\frac{(n-1)^2 \gamma_n}{1 - 2\lambda_n} \right]^{1/3}. \tag{3.113}$$

Putting $t = \max_i t_i^{(0)}$, we find from (3.113) that $t_i^{(0)} \leq t < 0.988 < 1$ for $3 \leq n \leq 13$, and $t_i^{(0)} \leq t < 0.999 < 1$ for $n \geq 14$, for each $i = 1, \ldots, n$. According to this, we infer from (3.112) that the sequences $\{t_i^{(m)}\}$ (and, consequently, $\{|u_i^{(m)}|\}$) tend to 0 for all $i = 1, \ldots, n$. Hence, the BSW method (3.107) is convergent.
    Putting $u^{(m)} = \max_{1 \leq i \leq n} |u_i^{(m)}|$, from (3.111), we get

$$u^{(m+1)} < \frac{\gamma_n}{(d^{(m)})^3} (n-1)^2 (u^{(m)})^4,$$

which means that the order of convergence of the BSW method is 4. □

The Halley-Like Method

Using a concept based on Bell's polynomials, X. Wang and Zheng [182] established a family of iterative methods of the order of convergence $k + 2$, where k is the highest order of the derivative of P appearing in the generalized iterative formula, see Sect. 1.1. For $k = 1$, this family gives the Ehrlich–Aberth's method (3.73), and for $k = 2$ produces the following iterative method of the fourth order for the simultaneous approximation of all simple zeros of a polynomial P

$$z_i^{(m+1)} = z_i^{(m)} - \cfrac{1}{f(z_i^{(m)}) - \cfrac{P(z_i^{(m)})}{2P'(z_i^{(m)})}\left(\left[S_{1,i}^{(m)}\right]^2 + S_{2,i}^{(m)}\right)} \quad (i \in \boldsymbol{I}_n,\ m=0,1,\dots),$$

(3.114)

where

$$f(z) = \frac{P'(z)}{P(z)} - \frac{P''(z)}{2P'(z)}, \quad S_{k,i}^{(m)} = \sum_{j \neq i} \frac{1}{\left(z_i^{(m)} - z_j^{(m)}\right)^k} \quad (k = 1, 2).$$

Since the function $f(z)$ appears in the well-known Halley's iterative method

$$\hat{z}_i = z_i - \frac{1}{\dfrac{P'(z_i)}{P(z_i)} - \dfrac{P''(z_i)}{2P'(z_i)}} = z_i - \frac{1}{f(z_i)},$$

we could say that the method (3.114) is of Halley's type. In the literature, the method (3.114) is sometimes called the Wang–Zheng's method.

The convergence analysis of the Halley-like method (3.114) is similar to that given previously in this section (see also the paper by M. Petković and Đ. Herceg [117]), so it will be presented in short.

Let us introduce the following abbreviations:

$$\rho_n = \frac{1}{1 - nc_n}, \quad \eta_n = \frac{2(1 - n\rho_n c_n)}{1 - \rho_n c_n} - \frac{n(n-1)(\rho_n c_n)^3(2 - \rho_n c_n)}{(1 - \rho_n c_n)^2},$$

$$\lambda_n = \frac{2\rho_n c_n(1 - \rho_n c_n + (n-1)\rho_n c_n)}{(1 - \rho_n c_n)\eta_n}, \quad \gamma_n = \frac{n(2 - \rho_n c_n)}{\eta_n(1 - \rho_n c_n)^2}.$$

Lemma 3.20. *Let $\hat{z}_1, \dots, \hat{z}_n$ be approximations generated by the iterative method (3.114) and let $\hat{u}_i = \hat{z}_i - \zeta_i$, $\hat{d} = \min\limits_{i \neq j} |\hat{z}_i - \hat{z}_j|$, and $\hat{w} = \max\limits_{1 \leq i \leq n} |\widehat{W}_i|$. If the inequality*

$$w < c_n d, \qquad c_n = \begin{cases} \dfrac{1}{3n + 2.4}, & 3 \leq n \leq 20 \\[2mm] \dfrac{1}{3n}, & n \geq 21 \end{cases}$$

(3.115)

holds, then:

(i) $d < \dfrac{1}{1 - 2\lambda_n}\hat{d}.$

(ii) $|u_i| < \rho_n c_n d.$

(iii) $\hat{w} < c_n \hat{d}.$

(iv) $|\hat{u}_i| \leq \dfrac{\gamma_n}{d^3}|u_i|^3 \displaystyle\sum_{j \neq i}|u_j|.$

The proof of this lemma is similar to the proofs of Lemmas 3.15 and 3.18.

We now give the convergence theorem for the iterative method (3.114) under computationally verifiable initial conditions.

Theorem 3.12. *Let P be a polynomial of degree $n \geq 3$ with simple zeros. If the initial condition*

$$w^{(0)} < c_n d^{(0)} \tag{3.116}$$

holds, where c_n is given by (3.115), then the Halley-like method (3.114) is convergent with the fourth order of convergence.

Proof. The proof of this theorem goes in a similar way to the previous cases using the assertions of Lemma 3.20. By virtue of the implication (iii) of Lemma 3.20 (i.e., $w < c_n d \Rightarrow \hat{w} < c_n \hat{d}$), we conclude by induction that the initial condition (3.116) implies the inequality $w^{(m)} < c_n d^{(m)}$ for each $m = 1, 2, \ldots$. For this reason, the assertions of Lemma 3.20 are valid for all $m \geq 0$. In particular (according to (i) and (iv) of Lemma 3.20), we have

$$\frac{d^{(m)}}{d^{(m+1)}} < \frac{1}{1 - 2\lambda_n} \tag{3.117}$$

and

$$|u_i^{(m+1)}| \leq \frac{\gamma_n}{(d^{(m)})^3}|u_i^{(m)}|^3 \sum_{\substack{j=1 \\ j \neq i}}^{n}|u_j^{(m)}| \quad (i \in \boldsymbol{I}_n) \tag{3.118}$$

for each iteration index $m = 0, 1, \ldots$.

Substituting

$$t_i^{(m)} = \left[\frac{(n-1)\gamma_n}{(1 - 2\lambda_n)(d^{(m)})^3}\right]^{1/3}|u_i^{(m)}|$$

into (3.118) gives

$$t_i^{(m+1)} \leq \frac{(1 - 2\lambda_n)d^{(m)}}{(n-1)d^{(m+1)}}\left(t_i^{(m)}\right)^3 \sum_{\substack{j=1 \\ j \neq i}}^{n} t_j^{(m)} \quad (i \in \boldsymbol{I}_n).$$

Hence, using (3.117), we arrive at

$$t_i^{(m+1)} < \frac{1}{n-1}\left(t_i^{(m)}\right)^3 \sum_{\substack{j=1 \\ j \neq i}}^{n} t_j^{(m)} \quad (i \in \boldsymbol{I}_n). \tag{3.119}$$

Since $|u_i^{(0)}| < \rho_n c_n d^{(0)}$ (assertion (ii) of Lemma 3.20), we may write

$$t_i^{(0)} = \left[\frac{(n-1)\gamma_n}{(1-2\lambda_n)(d^{(0)})^3}\right]^{1/3} |u_i^{(0)}| < \rho_n c_n \left[\frac{(n-1)\gamma_n}{1-2\lambda_n}\right]^{1/3}$$

for each $i = 1, \ldots, n$. Let $t_i^{(0)} \leq \max_i t_i^{(0)} = t$. Then

$$t < \rho_n c_n \left[\frac{(n-1)\gamma_n}{1-2\lambda_n}\right]^{1/3} < 0.310 \quad \text{for} \quad 3 \leq n \leq 20$$

and

$$t < 0.239 \qquad \text{for} \quad n \geq 21,$$

i.e., $t_i^{(0)} \leq t < 1$ for all $i = 1, \ldots, n$. Hence, we conclude from (3.119) that the sequences $\{t_i^{(m)}\}$ (and, consequently, $\{|u_i^{(m)}|\}$) tend to 0 for all $i = 1, \ldots, n$. Therefore, $z_i^{(m)} \to \zeta_i$ ($i \in \mathbf{I}_n$) and the method (3.114) is convergent.

Finally, from (3.118), there follows

$$|u_i^{(m+1)}| \leq u^{(m+1)} < (n-1)\frac{\gamma_n}{(d^{(m)})^3}\left(u^{(m)}\right)^4,$$

where $u^{(m)} = \max_{1 \leq i \leq n} |u_i^{(m)}|$. Therefore, the convergence order of the Halley-like method (3.114) is 4. □

Some Computational Aspects

In this section, we have improved the convergence conditions of four root finding methods. For the purpose of comparison, let us introduce the *normalized i-factor* $\Omega_n = n \cdot c_n$. The former Ω_n for the considered methods, found in the recent papers cited in Sect. 1.1, and the improved (new) Ω_n, proposed in this section, are given in Table 3.1.

Table 3.1 The entries of normalized i-factors

	Former Ω_n	New Ω_n
Ehrlich–Aberth's method (3.73)	$\dfrac{n}{2n+3}$	$\begin{cases} \dfrac{n}{2n+1.4} & (3 \leq n \leq 7), \\ 1/2 & (n \geq 8) \end{cases}$
EAN method (3.90)	$\dfrac{1}{3}$	$\begin{cases} \dfrac{n}{2.2n+1.9} & (3 \leq n \leq 21), \\ 1/2.2 & (n \geq 22) \end{cases}$
BSW method (3.107)	$\dfrac{n}{2n+2}$	$\begin{cases} \dfrac{n}{2n+1} & (3 \leq n \leq 13), \\ 1/2 & (n \geq 14) \end{cases}$
Halley-like method (3.114)	$\dfrac{1}{4}$	$\begin{cases} \dfrac{n}{3n+2.4} & (3 \leq n \leq 20), \\ 1/3 & (n \geq 21) \end{cases}$

To compare the former $\Omega_n = nc_n$ with the improved Ω_n, we introduce a percentage measure of the improvement

$$r\% = \frac{\Omega_n^{(\text{new})} - \Omega_n^{(\text{former})}}{\Omega_n^{(\text{former})}} \cdot 100.$$

Following Table 3.1, we calculated $r\%$ for $n \in [3, 30]$ and displayed $r\%$ in Fig. 3.3 as a function of n for each of the four considered methods. From Fig. 3.3, we observe that we have significantly improved i-factors c_n, especially for the EAN method (3.90) and Halley-like method (3.114).

The values of the i-factor c_n, given in the corresponding convergence theorems for the considered iterative methods, are mainly of theoretical importance. We were constrained to take smaller values of c_n to enable the validity

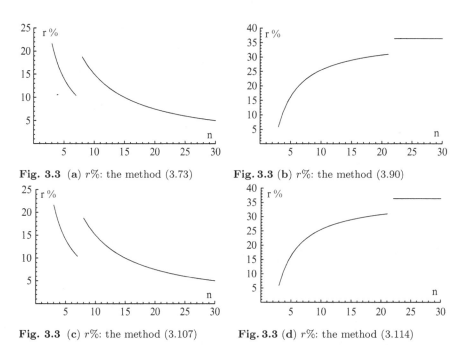

Fig. 3.3 (a) $r\%$: the method (3.73)   **Fig. 3.3** (b) $r\%$: the method (3.90)

Fig. 3.3 (c) $r\%$: the method (3.107)   **Fig. 3.3** (d) $r\%$: the method (3.114)

of inequalities appearing in the convergence analysis. However, these theoretical values of c_n can be suitably applied in ranking the considered methods regarding (1) their initial conditions for the guaranteed convergence and (2) convergence behavior in practice.

As mentioned in [118], in practical implementation of simultaneous root finding methods, we may take greater c_n related to that given in the convergence theorems and still preserve both guaranteed and fast convergence. The determination of the range of values of i-factor c_n providing favorable features (guaranteed and fast convergence) is a very difficult task, and practical experiments are the only means for obtaining some information on its range. We have tested the considered methods in examples of many algebraic polynomials with degree up to 20, taking initial approximations in such a way that the i-factor took the values kc_n for $k = 1$ (theoretical entry applied in the stated initial conditions) and for $k = 1.5$, 2, 3, 5, and 10. The stopping criterion was given by the inequality

$$\max_{1 \leq i \leq n} |z_i^{(m)} - \zeta_i| < 10^{-15}.$$

In Table 3.2, we give the average number of iterations (rounded to one decimal place), needed to satisfy this criterion.

From Table 3.2, we observe that the new i-factor not greater than $2c_n$ mainly preserves the convergence rate related to the theoretical value c_n given in the presented convergence theorems. The entry $3c_n$ is rather acceptable

Table 3.2 The average number of iterations as the i-factor increases

	c_n	$1.5c_n$	$2c_n$	$3c_n$	$5c_n$	$10c_n$
Ehrlich–Aberth's method (3.73)	3.9	4	4.2	5.4	7.3	13.3
EAN method (3.90)	3.1	3.2	3.4	5.1	6.1	10.2
BSW method (3.107)	3	3.1	3.3	4.3	5.8	9.8
Halley-like method (3.114)	3.2	3.4	4.2	5.5	6.7	10.7

from a practical point of view, while the choice of $5c_n$ doubles the number of iterations. Finally, the value $10c_n$ significantly decreases the convergence rate of all considered methods, although still provides the convergence.

3.4 A Posteriori Error Bound Methods

In this section, we combine good properties of iterative methods with fast convergence and a posteriori error bounds given in Corollary 1.1, based on Carstensen's results [13] on Gerschgorin's disks, to construct efficient inclusion methods for polynomial complex zeros. Simultaneous determination of both centers and radii leads to iterative error bound methods, which enjoy very convenient property of enclosing zeros at each iteration. This class of methods possesses a high computational efficiency since it requires less numerical operations compared with standard interval methods realized in interval arithmetic (see M. Petković and L. Petković [132]). Numerical experiments demonstrate equal or even better convergence behavior of these methods than the corresponding circular interval methods. In this section, the main attention is devoted to the construction of inclusion error bound methods. We will also give a review of some properties of these methods, including the convergence rate, efficient implementation, and initial conditions for the guaranteed convergence.

Corollary 1.1 given in Chap. 1 may be expressed in the following form.

Corollary 3.1. *Let P be an algebraic polynomial with simple (real or complex) zeros. Under the condition $w < c_n d$ ($c_n \leq 1/(2n)$), each of disks D_i defined by*

$$D_i = \left\{ z_i; \frac{|W_i(z_i)|}{1 - nc_n} \right\} = \{z_i; \rho_i\} \quad (i \in I_n)$$

contains exactly one zero of P.

If the centers z_i of disks D_i are calculated by an iterative method, then we can generate sequences of disks $D_i^{(m)}$ ($m = 0, 1, \ldots$) whose radii $\rho_i^{(m)} = W_i^{(m)}/(1 - nc_n)$ converge to 0 under some suitable conditions. It should be noted that only those methods which use quantities already calculated in the previous iterative step (in our case, the corrections W_i) enable a high computational efficiency. For this reason, we restrict our choice to the class of derivative-free methods which deal with Weierstrass' corrections, so-called *W-class*. The following most frequently used simultaneous methods from the *W*-class will be considered.

The Durand–Kerner's or Weierstrass' method [32], [72], shorter the W method, the convergence order 2:

$$z_i^{(m+1)} = z_i^{(m)} - W_i^{(m)} \quad (i \in I_n, \ m = 0, 1, \ldots). \tag{3.120}$$

The Börsch-Supan's method [10], shorter the BS method, the convergence order 3:

$$z_i^{(m+1)} = z_i^{(m)} - \frac{W_i^{(m)}}{1 + \displaystyle\sum_{j \neq i} \frac{W_j^{(m)}}{z_i^{(m)} - z_j^{(m)}}} \quad (i \in I_n, \ m = 0, 1, \ldots). \tag{3.121}$$

The Börsch-Supan's method with Weierstrass' correction [95], shorter the BSW method, the convergence order 4:

$$z_i^{(m+1)} = z_i^{(m)} - \frac{W_i^{(m)}}{1 + \displaystyle\sum_{j \neq i} \frac{W_j^{(m)}}{z_i^{(m)} - W_i^{(m)} - z_j^{(m)}}} \quad (i \in I_n, \ m = 0, 1, \ldots). \tag{3.122}$$

Let us note that $W_i^{(m)} = W(z_i^{(m)})$, see (1.17).

Let $z_1^{(0)}, \ldots, z_n^{(0)}$ be given initial approximations and let

$$z_i^{(m)} = \Phi_W(z_i^{(m-1)}) \quad (i \in I_n, \ m = 1, 2, \ldots) \tag{3.123}$$

be a derivative-free iterative method based on Weierstrass' corrections (belonging to the W-class), which is indicated by the subscript index "W." For example, the methods (3.120)–(3.122) belong to the W-class. Other iterative methods of Weierstrass' class are given in [34], [124], [131], [146], and [196].

Combining the results of Corollary 3.1 and (3.123), we can state the following inclusion method in a general form.

A posteriori error bound method. A posteriori error bound method (shorter PEB method) is defined by the sequences of disks $\{D_i^{(m)}\}$ ($i \in I_n$),

$$
D_i^{(0)} = \left\{ z_i^{(0)}; \frac{|W(z_i^{(0)})|}{1 - nc_n} \right\},
$$

$$
D_i^{(m)} = \{ z_i^{(m)}; \rho_i^{(m)} \}, \qquad (i \in I_n,\ m = 1, 2, \ldots), \qquad (3.124)
$$

$$
z_i^{(m)} = \Phi_W(z_i^{(m-1)}), \quad \rho_i^{(m)} = \frac{|W(z_i^{(m)})|}{1 - nc_n},
$$

assuming that the initial condition $w^{(0)} < c_n d^{(0)}$ (with $c_n \leq 1/(2n)$) holds.

The proposed method, defined by the sequences of disks given by (3.124), may be regarded as a quasi-interval method, which differs structurally from standard interval methods that deal with disks as arguments. For comparison, let us present the following circular interval methods which do not use the polynomial derivatives.

The Weierstrass-like interval method [183], the order 2:

$$
Z_i^{(m+1)} = z_i^{(m)} - \frac{P(z_i^{(m)})}{\displaystyle\prod_{\substack{j=1 \\ j \neq i}}^{n} (z_i^{(m)} - Z_j^{(m)})} \qquad (i \in I_n,\ m = 0, 1, \ldots). \qquad (3.125)
$$

The Börsch-Supan-like interval method [107], the order 3:

$$
Z_i^{(m+1)} = z_i^{(m)} - \frac{W_i^{(m)}}{1 + \displaystyle\sum_{\substack{j=1 \\ j \neq i}}^{n} \frac{W_j^{(m)}}{Z_i^{(m)} - z_j^{(m)}}} \qquad (i \in I_n,\ m = 0, 1, \ldots). \qquad (3.126)
$$

The Börsch-Supan-like interval method with Weierstrass' correction [111], the order 4 (with the centered inversion (1.63)):

$$
Z_i^{(m+1)} = z_i^{(m)} - \frac{W_i^{(m)}}{1 + \displaystyle\sum_{\substack{j=1 \\ j \neq i}}^{n} \frac{W_j^{(m)}}{Z_i^{(m)} - W_i^{(m)} - z_j^{(m)}}} \qquad (i \in I_n,\ m = 0, 1, \ldots).
$$

$$
(3.127)
$$

All of the methods (3.124)–(3.127) possess the crucial inclusion property: each of the produced disks contains exactly one zero in each iteration. In this manner, not only very close zero approximations (given by the centers of disks) but also the upper error bounds for the zeros (given by the radii of disks) are obtained. More about interval methods for solving polynomial equations can be found in the books by M. Petković [109] and M. Petković and L. Petković [129].

Studying the convergence of error bounds produced by (3.124), the following important tasks arise:

1. Determine the convergence order of a posteriori error bound method when the centers $z_i^{(m)}$ of disks

$$D_i^{(m)} = \left\{ z_i^{(m)}; \frac{|W(z_i^{(m)})|}{1 - nc_n} \right\} \quad (i \in \mathbf{I}_n, \ m = 0, 1, \dots) \qquad (3.128)$$

   are calculated by an iterative method of order $k \ (\geq 2)$.
2. State computationally verifiable initial condition that guarantees the convergence of the sequences of radii $\{\operatorname{rad} D_i^{(m)}\}$. We note that this problem, very important in the theory and practice of iterative processes in general, is a part of Smale's point estimation theory [165] which has attracted a considerable attention during the last two decades (see [118] and Chap. 2 for details). As mentioned in the previous sections, initial conditions in the case of algebraic polynomials should depend only on attainable data – initial approximations, polynomial degree, and polynomial coefficients.
3. Compare the computational efficiencies of the PEB methods and the existing circular interval methods (given, for instance, by (3.125)–(3.127)). Which of these two classes of methods is more efficient?
4. Using numerical experiments, compare the size of inclusion disks produced by the PEB methods and the corresponding interval methods (3.125)–(3.127). Whether the construction of PEB methods is justified?

The study of these subjects was the main goal of the paper [122]. Here, we give the final results and conclusions of this investigation in short.

Assume that the following inequality

$$w^{(0)} < c_n d^{(0)} \qquad (3.129)$$

holds, where c_n is given by

$$c_n = \begin{cases} \dfrac{1}{2n}, & \text{the W method [2] and BS method [42],} \\[3mm] \dfrac{1}{2n+1}, & \text{the BSW method [52].} \end{cases} \qquad (3.130)$$

Then, the following three methods from the W-class are convergent: the Durand–Kerner's method (3.120) (for the proof, see Batra [5]), the Börsch-Supan's method (3.121) (M. Petković and Đ. Herceg [117]), and the Börsch-Supan's method with Weierstrass' correction (3.122) (see [60], [140]). The corresponding inequalities of the form

$$|W_i^{(m+1)}| < \delta_n |W_i^{(m)}| \quad (\delta_n < 1)$$

are the composite parts of Lemmas 3.3(i), 3.6(i), and 3.19(i) under the condition (3.129) for specific entries c_n given by (3.130). This means that the sequences $\{|W_i^{(m)}|\}$ ($i \in I_n$) are convergent and tend to 0. Hence, the sequences of radii $\{\rho^{(m)}|\}$ ($i \in I_n$) are also convergent and tend to 0 under the condition (3.129). The convergence rate of the PEB methods based on the iterative methods (3.120)–(3.122) was studied in [122], where the following assertions were proved.

Theorem 3.13. *The PEB method (3.124), based on the Durand–Kerner's method (3.120), converges quadratically if the initial condition (3.129) holds, where $c_n = 1/(2n)$.*

Theorem 3.14. *The PEB method (3.124), based on the Börsch-Supan's method (3.121), converges cubically if the initial condition (3.129) holds, where $c_n = 1/(2n)$.*

Theorem 3.15. *The PEB method (3.124), based on the Börsch-Supan's method with Weierstrass' corrections (3.122), converges with the order 4 if the initial condition (3.129) holds, where $c_n = 1/(2n+1)$.*

We emphasize that the initial condition (3.129) (with c_n given by (3.130)) that guarantees the convergence of the PEB methods (3.124)–(3.120), (3.124)–(3.121), and (3.124)–(3.122) depends only on attainable data, which is of great practical importance.

Computational Aspects

In the continuation of this section, we give some practical aspects in the implementation of the proposed methods. As mentioned above, the computational cost significantly decreases if the quantities $W_i^{(0)}, W_i^{(1)}, \ldots$ ($i \in I_n$), necessary in the calculation of the radii $\rho_i^{(m)} = |W_i^{(m)}|/(1 - nc_n)$, are applied in the calculation of the centers $z_i^{(m+1)}$ defined by the employed iterative formula from the W-class. Regarding the iterative formulae (3.120)–(3.122), we observe that this requirement is satisfied. A general calculating procedure can be described by the following algorithm.

Calculating Procedure (I)

Given $z_1^{(0)}, \ldots, z_n^{(0)}$ and the tolerance parameter τ.
Set $m = 0$.
1° *Calculate Weierstrass' corrections $W_1^{(m)}, \ldots, W_n^{(m)}$ at the points $z_1^{(m)},$*
$\ldots, z_n^{(m)}$.
2° *Calculate the radii $\rho_i^{(m)} = |W_i^{(m)}|/(1 - nc_n)$ $(i = 1, \ldots, n)$.*
3° *If $\max\limits_{1 \le i \le n} \rho_i^{(m)} < \tau$, then STOP*
   otherwise, GO TO 4°.
4° *Calculate the new approximations $z_1^{(m+1)}, \ldots, z_n^{(m+1)}$ by a suitable iterative*
   formula from the W-class (for instance, by (3.120), (3.121), or (3.122)).
5° *Set $m := m + 1$ and GO TO the step 1°.*

Following the procedure (I), we have realized many numerical examples and, for demonstration, we select the following one.

Example 3.1. We considered the polynomial

$$P(z) = z^{13} - (5 + 5i)z^{12} + (5 + 25i)z^{11} + (15 - 55i)z^{10} - (66 - 75i)z^9$$
$$+ 90z^8 - z^5 + (5 + 5i)z^4 - (5 + 25i)z^3 - (15 - 55i)z^2$$
$$+ (66 - 75i)z - 90$$
$$= (z - 3)(z^8 - 1)(z^2 - 2z + 5)(z - 2i)(z - 3i).$$

Starting from sufficiently close initial approximations $z_1^{(0)}, \ldots, z_{13}^{(0)}$, we first calculated the radii $\rho_i^{(0)} = |W(z_i^{(0)}|/(1 - nc_n)$ of initial disks $D_1^{(0)}, \ldots, D_{13}^{(0)}$. These disks were applied in the implementation of a posteriori error bound methods (3.124) as well as interval methods (3.125)–(3.127). We obtained $\max \rho_i^{(0)} = 0.3961$ for the methods (3.125), (I-W), (3.126), (I-BS) and $\max \rho_i^{(0)} = 0.3819$ for (3.127) and (I-BSW). The approximations $z_i^{(m)}$ $(m \ge 1)$ were calculated by the iterative formulae (3.120)–(3.122) and the corresponding inclusion methods are referred to as (I-W), (I-BS), and (I-BSW), respectively. The largest radii of the disks obtained in the first four iterations may be found in Table 3.3, where $A(-q)$ means $A \times 10^{-q}$.

Table 3.3 Resulting disks obtained by Procedure I

Methods	$\max \rho_i^{(1)}$	$\max \rho_i^{(2)}$	$\max \rho_i^{(3)}$	$\max \rho_i^{(4)}$
(I-W) (3.124)–(3.120)	1.26(−1)	1.74(−2)	1.33(−4)	1.59(−8)
Interval W (3.125)	1.05	No inclusions	–	–
(I-BS) (3.124)–(3.121)	1.83(−2)	3.61(−6)	1.32(−17)	7.06(−52)
Interval BS (3.126)	1.99(−1)	2.41(−4)	2.39(−15)	2.38(−49)
(I-BSW) (3.124)–(3.122)	6.94(−3)	4.92(−10)	1.83(−38)	3.20(−152)
Interval BSW (3.127)	2.98(−1)	1.47(−5)	1.81(−24)	2.68(−100)

In our calculation, we employed multiprecision arithmetic in *Mathematica* 6.0 since the tested methods converge extremely fast producing very small disks. From Table 3.3, we observe that the PEB methods are equal or better than the corresponding methods (of the same order) (3.125)–(3.127) realized in complex interval arithmetic. A number of numerical experiments showed similar convergence behavior of the tested methods.

The Weierstrass' interval method (3.125) exhibits rather poor results. The explanation lies in the fact that this method uses the product of disks which is not an exact operation in circular arithmetic and produces oversized disks (see Sect. 1.3).

Calculation Procedure (I) assumes the knowledge of initial approximations $z_1^{(0)}, \ldots, z_n^{(0)}$ in advance. The determination of these approximations is usually realized by a slowly convergent multistage composite algorithm. Sometimes, the following simple approach gives good results in practice.

Calculating Procedure (II)

$1°$ *Find the disk centered at the origin with the radius*

$$R = 2 \max_{1 \le k \le n} \left| a_{n-k} \right|^{1/k} \quad (see\ (1.58)\ or\ (5.72)),$$

which contains all zeros of the polynomial $P(z) = z^n + a_{n-1} z^{n-1} + \cdots + a_1 z + a_0$.

$2°$ *Calculate Aberth's initial approximations* [1]

$$z_\nu^{(0)} = -\frac{a_{n-1}}{n} + r_0 \exp(\mathrm{i}\theta_\nu), \quad \mathrm{i} = \sqrt{-1}, \quad \theta_\nu = \frac{\pi}{n}\left(2\nu - \frac{3}{2}\right) \quad (\nu = 1, \ldots, n),$$

equidistantly distributed along the circle $|z + a_{n-1}/n| = r_0$, $r_0 \le R$ *(see Sect. 4.4)*.

$3°$ *Apply the simultaneous method* (3.120) *or* (3.121) *starting with Aberth's approximations; stop the iterative process when the condition*

$$\max_{1 \le i \le n} |W(z_i^{(m)})| < c_n \min_{i \ne j} |z_i^{(m)} - z_j^{(m)}| \tag{3.131}$$

is satisfied.

$4°$–$8°$ *The same as the steps* $1°$–$5°$ *of Procedure I.*

We applied Procedure II on the following example.

Example 3.2. To find approximations to the zeros of the polynomial

$$z^{15} + z^{14} + 1 = 0$$

satisfying the condition (3.131) (with $c_n = 1/(2n)$), we applied the Börsch-Supan's method (3.121) with Aberth's initial approximations taking $a_{n-1} = 1$, $n = 15$, and $r_0 = 2$. The condition (3.131) was satisfied after seven iterative steps. The obtained approximations were used to start the PEB methods (I-W), (I-BS), and (I-BSW). After three iterations, we obtained disks whose largest radii are given in Table 3.4.

Table 3.4 Resulting disks obtained by (I-W), (I-BS), and (I-BSW): Procedure II

Methods	$\max \rho_i^{(0)}$	$\max \rho_i^{(1)}$	$\max \rho_i^{(2)}$
(I-W) (3.124)–(3.120)	$1.51(-3)$	$3.79(-6)$	$2.27(-11)$
(I-BS) (3.124)–(3.121)	$1.51(-3)$	$4.10(-9)$	$8.31(-26)$
(I-BSW) (3.124)–(3.122)	$1.46(-3)$	$9.64(-12)$	$1.60(-44)$

From Tables 3.3 and 3.4, we observe that the results obtained by the methods (I-W), (I-BS), and (I-BSW) coincide with the theoretical results given in Corollary 3.1 and Theorems 3.13–3.15; in other words, the order of convergence in practice matches very well the order expressed in Theorems 3.13–3.15.

At the beginning of the section, we mentioned that the PEB methods require less numerical operations compared with their counterparts in complex interval arithmetic. In Table 3.5, we give the total number of numerical operations per one iteration, reduced to real arithmetic operations. We have used the following abbreviations:

- $AS(n)$ (total number of additions and subtractions)
- $M(n)$ (multiplications)
- $D(n)$ (divisions)

Table 3.5 The number of basic operations

	$AS(n)$	$M(n)$	$D(n)$
(I-W) (3.124)–(3.120)	$8n^2 + n$	$8n^2 + 2n$	$2n$
Interval W (3.125)	$22n^2 - 6n$	$25n^2 - 6n$	$8n^2 - n$
(I-BS) (3.124)–(3.121)	$15n^2 - 6n$	$14n^2 + 2n$	$2n^2 + 2n$
Interval BS (3.126)	$23n^2 - 4n$	$23n^2 + 2n$	$7n^2 + 2n$
(I-BSW) (3.124)–(3.122)	$15n^2 - 4n$	$14n^2 + 2n$	$2n^2 + 2n$
Interval BSW (3.127)	$23n^2 - 2n$	$23n^2 + 2n$	$7n^2 + 2n$

From Table 3.5, we observe that the PEB methods require significantly less numerical operations with respect to the corresponding interval methods. One of the reasons for this advantage is the use of the already calculated Weierstrass' corrections W_i in the evaluation of the radii ρ_i.

Parallel Implementation

It is worth noting that the error bound method (3.124) for the simultaneous determination of all zeros of a polynomial is very suitable for the implementation on parallel computers since it runs in several identical versions. In this manner, a great deal of computation can be executed simultaneously. An analysis of total running time of a parallel iteration and the determination of the optimal number of processors points to some undoubted advantages of the implementation of simultaneous methods on parallel processing computers, see, e.g., [22]–[24], [44], [115]. The parallel processing becomes of great interest to speed up the determination of zeros when one should treat polynomials with degree 100 and higher, appearing in mathematical models in scientific engineering, including digital signal processing or automatic control [66], [92].

The model of parallel implementation is as follows: It is assumed that the number of processors k $(\leq n)$ is given in advance. Let

$$\boldsymbol{W}^{(m)} = \left(W_1^{(m)}, \ldots, W_n^{(m)}\right),$$
$$\boldsymbol{\rho}^{(m)} = \left(\rho_1^{(m)}, \ldots, \rho_n^{(m)}\right),$$
$$\boldsymbol{z}^{(m)} = \left(z_1^{(m)}, \ldots, z_n^{(m)}\right)$$

denote vectors at the mth iterative step, where $\rho_i^{(m)} = |W(z_i^{(m)})|/(1 - nc_n)$, and $z_i^{(m)}$ is obtained by the iterative formula $z_i^{(m)} = \Phi_W(z_i^{(m-1)})$ $(i \in \boldsymbol{I}_n)$. The starting vector $\boldsymbol{z}^{(0)}$ is computed by all processors $C_1, \ldots, C_k$ using some suitable globally convergent method based on a subdivided procedure and the inclusion annulus $\{z : r \leq |z| \leq R\}$ which contains all zeros, given later by (4.72).

In the next stage, each step consists of sharing the calculation of $W_i^{(m)}$, $\rho_i^{(m)}$, and $z_i^{(m+1)}$ among the processors and in updating their data through a broadcast procedure (shorter $BCAST(\boldsymbol{W}^{(m)}, \boldsymbol{\rho}^{(m)})$, $BCAST(\boldsymbol{z}^{(m+1)})$). As in [23], let $I_1, \ldots, I_k$ be disjunctive partitions of the set $\{1, \ldots, n\}$, where $\cup I_j = \{1, \ldots, n\}$. To provide good load balancing between the processors, the index sets $I_1, \ldots, I_k$ are chosen so that the number of their components $\mathbb{N}(I_j)$ $(j = 1, \ldots, k)$ is determined as $\mathbb{N}(I_j) \leq \left[\frac{n}{k}\right]$. At the mth iterative step, the processor C_j $(j = 1, \ldots, k)$ computes $W_i^{(m)}, \rho_i^{(m)}$, and, if necessary, $z_i^{(m+1)}$ for all $i \in I_j$ and then it transmits these values to all other processors using a broadcast procedure. The program terminates when the stopping criterion is satisfied, say, if for a given tolerance τ the inequality

$$\max_{1 \leq i \leq n} \left|\rho_i^{(m)}\right| < \tau$$

holds. A program written in pseudocode for a parallel implementation of the error bound method (3.124) is given below.

Program A POSTERIORI ERROR BOUND METHOD
begin
      for all $j = 1, \ldots, k$ **do** determination of the approximations $\boldsymbol{z}^{(0)}$;
      $m := 0$
      $C :=$false

do

     for all $j = 1, \ldots, k$ **do in parallel**

     begin

          Compute $W_i^{(m)}$, $i \in I_j$;

          Compute $\rho_i^{(m)}$, $i \in I_j$;

          Communication: $\mathrm{BCAST}\big(\boldsymbol{W}^{(m)}, \boldsymbol{\rho}^{(m)}\big)$;

     end

     if $\max\limits_{1 \leq i \leq n} \rho_i^{(m)} < \tau$; $C :=$true

     else

          $m := m + 1$

          for all $j = 1, \ldots, k$ **do in parallel**

          begin

               Compute $z_i^{(m)}$, $i \in I_j$, by (3.123);

               Communication: $\mathrm{BCAST}\big(\boldsymbol{z}^{(m)}\big)$;

          end

     endif

until C

OUTPUT $\boldsymbol{z}^{(m)}, \boldsymbol{\rho}^{(m)}$

end

Chapter 4
Families of Simultaneous Methods of Higher Order: Part I

The aim of this chapter is to present the fourth-order families of simultaneous methods for the determination of polynomial (simple or multiple) zeros. These methods are based on Hansen–Patrick's one-parameter family [55] with the cubic convergence (Sect. 4.1). First, we present the derivative-free family of methods for the simultaneous approximation of simple zeros and show that the methods of this family have the convergence order equal to 4 (Sect. 4.2). Next, we give computationally verifiable initial conditions that provide the guaranteed convergence and state the convergence theorem. In the second part of this chapter (Sect. 4.3), we study another family of methods that uses derivatives, also based on Hansen–Patrick's formula. Aside from convergence analysis, we construct families of methods for finding multiple zeros of a polynomial. To demonstrate the convergence speed of the considered families, several numerical examples are included at the end of the chapter (Sect. 4.4).

4.1 Hansen–Patrick's Family of Root Finding Methods

Let f be an analytic function in a complex domain with a simple or multiple zero ζ. The problem of extracting zeros is extensively investigated in the literature on this subject and many efficient iterative methods have been developed. Among them, the third-order methods such as Euler's, Laguerre's, Halley's, and Ostrowski's method have an important role. Such a wide range of methods opens the question of their mutual dependance and the equivalency of some methods.

One attempt to unify the class of methods with cubic convergence was presented in the paper [55] of Hansen and Patrick by the family

$$\hat{z} = z - \frac{(\alpha + 1)f(z)}{\alpha f'(z) \pm \sqrt{f'(z)^2 - (\alpha + 1)f(z)f''(z)}}, \tag{4.1}$$

M. Petković, *Point Estimation of Root Finding Methods*. Lecture Notes in Mathematics 1933,
© Springer-Verlag Berlin Heidelberg 2008

where α ($\neq -1$) is a parameter and $\hat{z}$ denotes a new approximation. This family has a cubic convergence for a finite α (for the proof, see [55]).

Starting from the Hansen–Patrick's iterative formula (4.1), some well-known methods are obtained by a suitable choice of the parameter α. We illustrate this by several examples:

1. Taking $\alpha = 0$, (4.1) reduces to *Ostrowski's method* [113]

$$\hat{z} = z - \frac{f(z)}{\sqrt{f'(z)^2 - f''(z)f(z)}}. \tag{4.2}$$

2. Setting $\alpha = 1$ in (4.1), we directly obtain the third-order *Euler's method* [146], [172]

$$\hat{z} = z - \frac{2f(z)}{f'(z) \pm \sqrt{f'(z)^2 - 2f(z)f''(z)}}, \tag{4.3}$$

   sometimes called Halley's irrational method [4], [45].

3. For $\alpha = -1$, we apply a limiting process and generate *Halley's method* [4], [45]

$$\hat{z} = z - \frac{f(z)}{f'(z) - \dfrac{f(z)f''(z)}{2f'(z)}}. \tag{4.4}$$

4. Taking $\alpha = 1/(\nu - 1)$ in (4.1), we obtain *Laguerre's method*

$$\hat{z} = z - \frac{\nu f(z)}{f'(z) \pm \sqrt{(\nu - 1)^2 f'(z)^2 - \nu(\nu - 1)f(z)f''(z)}}, \tag{4.5}$$

   where ν ($\neq 0, 1$) is a parameter. Extensive studies of Laguerre's method (4.2) can be found in [99] (see also [8], [55], [84], [102]). Two modifications of Laguerre's method, which enable simultaneous determination of all simple zeros of a polynomial and have the order of convergence at least 4, were presented by Hansen, Patrick, and Rusnak [56]. Further improvements of these methods were proposed in [134].

5. If we let $\alpha \to \infty$ in (4.1), we obtain *Newton's method*

$$\hat{z} = z - \frac{f(z)}{f'(z)}, \tag{4.6}$$

   which has the quadratic convergence.

Remark 4.1. In [55], Hansen and Patrick derived a family of zero-finding methods (4.1) through an extensive procedure. Actually, this family is not new; as shown in [135], it can be obtained from Laguerre's method (4.5) by a special choice of the parameter ν. Indeed, substituting $\nu = 1/\alpha + 1$ in (4.5), we obtain Hansen–Patrick's formula (4.1).

Remark 4.2. According to Henrici [57, p. 532], the argument of the square root appearing in the above iterative formulae *is to be chosen to differ by less than $\pi/2$ from the argument of $f'(z)$.*

Remark 4.3. It is known that Laguerre's and Halley's method converge globally and monotonically in the case when f is a polynomial with all real roots (see, e.g., Davies and Dawson [27]). Besides, Laguerre's method shows extremely good behavior when $|z|$ is large, see Parlett [102].

Multiple Zeros

Let ζ be the zero of f of the (known) multiplicity μ. Hansen and Patrick have started from the function f in the form $f(z) = (z - \zeta)^\mu g(z)$ $(g(\zeta) \neq 0)$ and obtained a one-parameter family of methods for finding a multiple zero,

$$\hat{z} = z - \frac{\mu(\mu\alpha + 1)f(z)}{\mu\alpha f'(z) \pm \sqrt{\mu(\alpha(\mu - 1) + 1)f'(z)^2 - \mu(\mu\alpha + 1)f(z)f''(z)}}. \quad (4.7)$$

Let us consider some special cases corresponding to those given for simple zeros (see (1.41)–(1.45)):

1. If we set $\alpha = 0$ in (4.7), then we obtain the well-known *Ostrowski's method* of the third order

$$\hat{z} = z - \frac{\sqrt{\mu}f(z)}{\sqrt{f'(z)^2 - f(z)f''(z)}}. \quad (4.8)$$

2. Letting $\alpha = 1/\mu$ in (4.7), we obtain the third-order *Euler's method* for multiple zeros

$$\hat{z} = z - \frac{2\mu f(z)}{f'(z) \pm \sqrt{(2\mu - 1)f'(z)^2 - 2\mu f(z)f''(z)}}. \quad (4.9)$$

3. Taking $\alpha = -1/\mu$ and applying a limiting process in (4.7), one obtains

$$\hat{z} = z - \frac{f(z)}{\dfrac{\mu + 1}{2\mu}f'(z) - \dfrac{f(z)f''(z)}{2f'(z)}}. \quad (4.10)$$

This is *Halley's method* for multiple zeros of the third order.

4. Putting $\alpha = 1/(\nu - \mu)$ in (4.7), we get the counterpart of Laguerre's method

$$\hat{z} = z - \frac{\nu f(z)}{f'(z) \pm \sqrt{\left(\dfrac{\nu - \mu}{\mu}\right)\left[(\nu - 1)f'(z)^2 - \nu f(z)f''(z)\right]}}, \quad (4.11)$$

which was known to Bodewig [8].

5. Letting $\alpha \to \infty$ in (4.7), we obtain the second-order *Newton's method* for multiple zeros, known also as Schröder's method [160],

$$\hat{z} = z - \mu \frac{f(z)}{f'(z)}. \tag{4.12}$$

Remark 4.4. Let us introduce the function

$$F(z) = f(z)^{1/\mu}$$

for which ζ is a simple zero. Applying Hansen–Patrick's formula (4.1) to the function F, we obtain the iterative process for finding a multiple zero [135]

$$\hat{z} = z - \frac{\mu(\alpha + 1)f(z)}{\alpha f'(z) \pm \sqrt{(\mu(\alpha + 1) - \alpha)f'(z)^2 - \mu(\alpha + 1)f(z)f''(z)}}. \tag{4.13}$$

This is the simplified version of Hansen–Patrick's formula (4.7); indeed, formally replacing α with $\mu\alpha$ in (4.13), we obtain (4.7).

4.2 Derivative-Free Family of Simultaneous Methods

We consider now a special case when the function f is an algebraic polynomial. Let P be a monic polynomial of degree n with simple zeros $\zeta_1, \ldots, \zeta_n$ and let $z_1, \ldots, z_n$ be n pairwise distinct approximations to these zeros. In Sect. 1.1, we have dealt with the Weierstrass' correction

$$W_i = \frac{P(z_i)}{\displaystyle\prod_{\substack{j=1 \\ j \neq i}}^{n} (z_i - z_j)} \quad (i \in \boldsymbol{I}_n). \tag{4.14}$$

Using Weierstrass' corrections $W_1, \ldots, W_n$ and approximations $z_1, \ldots, z_n$, by the Lagrangean interpolation, we can represent the polynomial P for all $z \in \mathbb{C}$ as

$$P(z) = \prod_{j=1}^{n} (z - z_j) + \sum_{k=1}^{n} W_k \prod_{\substack{j=1 \\ j \neq k}}^{n} (z - z_j). \tag{4.15}$$

Recall that the Weierstrass' function $W_i(z)$ has been introduced in Sect. 1.1 by (1.17), i.e.,

$$W_i(z) = \frac{P(z)}{\displaystyle\prod_{\substack{j=1 \\ j \neq i}}^{n} (z - z_j)}.$$

Using (4.15), we get (compare with $F(2)$ in (1.17))

$$W_i(z) = W_i + (z - z_i)\left(1 + \sum_{\substack{j=1 \\ j \neq i}}^{n} \frac{W_j}{z - z_j}\right). \tag{4.16}$$

Note that any zero ζ_i of P is also a zero of the function $W_i(z)$.

For simplicity, we use the corrections W_i and introduce the following abbreviations:

$$G_{1,i} = \sum_{\substack{j=1 \\ j \neq i}}^{n} \frac{W_j}{z_i - z_j}, \quad G_{2,i} = \sum_{\substack{j=1 \\ j \neq i}}^{n} \frac{W_j}{(z_i - z_j)^2}.$$

Starting from (4.16), we find

$$W_i(z_i) = W_i, \quad W_i'(z_i) = 1 + \sum_{\substack{j=1 \\ j \neq i}}^{n} \frac{W_j}{z_i - z_j} = 1 + G_{1,i},$$

$$W_i''(z_i) = -2 \sum_{\substack{j=1 \\ j \neq i}}^{n} \frac{W_j}{(z_i - z_j)^2} = -2G_{2,i}. \tag{4.17}$$

Applying Hansen–Patrick's formula (4.1) to the function $W_i(z)$ given by (4.16), M. Petković et al. [124] derived the following one-parameter family for the simultaneous approximation of all simple zeros of a polynomial P:

$$\hat{z}_i = z_i - \frac{(\alpha + 1)W_i}{\alpha(1 + G_{1,i}) \pm \sqrt{(1 + G_{1,i})^2 + 2(\alpha + 1)W_i G_{2,i}}} \quad (i \in \boldsymbol{I}_n). \tag{4.18}$$

We will prove later that the order of convergence of the iterative methods of the family (4.18) is equal to 4 for any fixed and finite parameter α. This family of methods enables (1) simultaneous determination of all zeros of a given polynomial and (2) the acceleration of the order of convergence from *3* to *4*. A number of numerical experiments have shown that the proposed methods possess very good convergence properties.

Now, we present some special cases of the iterative formula (4.18).

For $\alpha = 0$, the family (4.18) gives the *Ostrowski-like method*

$$\hat{z}_i = z_i - \frac{W_i}{\sqrt{(1 + G_{1,i})^2 + 2W_i G_{2,i}}} \quad (i \in \boldsymbol{I}_n). \tag{4.19}$$

As in the case of other considered methods, the name comes from the fact that the method (4.19) can be obtained by applying the Ostrowski's method (4.2) to the function $W_i(z)$.

Setting $\alpha = 1$ in (4.18), we obtain the *Euler-like method*

$$\hat{z}_i = z_i - \frac{2W_i}{1 + G_{1,i} \pm \sqrt{(1 + G_{1,i})^2 + 4W_i G_{2,i}}} \quad (i \in \boldsymbol{I}_n). \tag{4.20}$$

If we let $\alpha = 1/(n-1)$, where n is the polynomial degree, (4.18) becomes the *Laguerre-like method*

$$\hat{z}_i = z_i - \frac{nW_i}{1 + G_{1,i} \pm \sqrt{((n-1)(1+G_{1,i}))^2 + 2n(n-1)W_iG_{2,i}}} \quad (i \in \boldsymbol{I}_n).$$

(4.21)

The case $\alpha = -1$ is not obvious at first sight and it requires a limiting operation in (4.18). After short calculation, we find that $\alpha = -1$ yields

$$\hat{z}_i = z_i - \frac{W_i(1+G_{1,i})}{(1+G_{1,i})^2 + W_iG_{2,i}} \quad (i \in \boldsymbol{I}_n).$$

(4.22)

This formula can be derived directly by applying the classical Halley's formula (4.4) to the Weierstrass' function $W_i(z)$, so that (4.22) will be referred to as the *Halley-like method*. Let us note that Ellis and Watson [34] derived the iterative formula (4.22) using a different approach.

Letting $\alpha \to \infty$ in (4.18), we get

$$\hat{z}_i = z_i - \frac{W_i}{1 + G_{1,i}} = z_i - \frac{W_i}{1 + \displaystyle\sum_{j \neq i} \frac{W_j}{z_i - z_j}} \quad (i \in \boldsymbol{I}_n).$$

(4.23)

This is the third-order iterative method proposed for the first time by Börsch-Supan [10]. Let us note that this method can be directly obtained by applying Newton's method (4.6) to the Weierstrass' function $W_i(z)$.

Note that the iterative formula (4.18) contains a "$\pm$" in front of the square root. We should choose the sign in such a way that the denominator is larger in magnitude. If approximations are reasonably close to the zeros of P, which is the case when we deal with the initial conditions considered in this book, then a simple analysis similar to that presented in [174] shows that we should take the sign "+." Such a choice ensures that the main part of the iterative formula (4.18) is the cubically convergent Börsch-Supan's method (4.23). In the case of the minus sign, the iterative formula (4.18) behaves as

$$\hat{z}_i = z_i - \frac{\alpha+1}{\alpha-1} \cdot \frac{W_i}{1+G_{1,i}} \quad (i \in \boldsymbol{I}_n),$$

which gives only a linearly convergent method for a finite α. We must therefore take the plus sign, i.e., we write (4.18) in the form

$$\hat{z}_i = z_i - \frac{(\alpha+1)W_i}{(1+G_{1,i})(\alpha + \sqrt{1 + 2(\alpha+1)t_i}\,)} \quad (i \in \boldsymbol{I}_n),$$

(4.24)

where

$$t_i = \frac{W_iG_{2,i}}{(1+G_{1,i})^2}.$$

If $\alpha = -1$, then applying a limiting operation in (4.24) we obtain the Halley-like simultaneous method (see (4.22))

$$\hat{z}_i = z_i - \frac{W_i}{(1 + G_{1,i})(1 + t_i)} \quad (i \in \boldsymbol{I}_n). \tag{4.25}$$

Remark 4.5. In this section, we consider iterative methods from the family (4.18) with the order of convergence 4. For this reason, we will assume that the parameter α is not too large in magnitude. The convergence analysis presented in [124] shows that large values of $|\alpha|$ give the methods whose convergence rate decreases and approaches 3. In practice, in such situations, the iterative formula (4.18) behaves as the aforementioned cubically convergent Börsch-Supan's method (4.23). In that case, the initial convergence conditions may be weakened, see [114] and Lemma 3.7.

Let us introduce the denotations

$$q_n := \frac{(n - 1)c_n}{(1 - (n - 1)c_n)^2}, \quad h_n := \sqrt{1 - 2|\alpha + 1|q_n}.$$

The parameter α can generally take entries from the complex-valued set. However, such a choice does not yield any advantages so that, in practice, we deal with a real α for simplicity. In this analysis, h_n must be a real nonnegative quantity. For this reason, we start from the inequality $2|\alpha + 1|q_n \leq 1$ and obtain the following range for the parameter α, called the α-*disk*:

$$A_\alpha(n) := \left\{ -1; \frac{1}{2q_n} \right\} \supseteq \left\{ -1; \frac{2c_3^2}{(1 - 2c_3)^2} \right\}, \quad n \geq 3.$$

Order of Convergence

Now, we prove that the iterative methods of the family (4.18) have the order of convergence equals 4 for any fixed and finite parameter α. In our convergence analysis, we will use the notation introduced above. Besides, let $\hat{u}_i = \hat{z}_i - \zeta_i$ and $u_i = z_i - \zeta_i$ be the errors in two successive iterations. For any two complex numbers z and w which are of the same order in magnitude, we will write $z = \mathcal{O}_M(w)$. In our analysis, we will suppose that the errors $u_1, \ldots, u_n$ are of the same order in magnitude, i.e., $u_i = \mathcal{O}_M(u_j)$ for any pair $i, j \in \boldsymbol{I}_n$. Furthermore, let $u_* \in \{u_1, \ldots, u_n\}$ be the error with the maximal magnitude (i.e., $|u_*| \geq |u_i|$ $(i = 1, \ldots, n)$ but still $u_* = \mathcal{O}_M(u_i)$ for any $i \in \boldsymbol{I}_n$).

Theorem 4.1. *If the approximations $z_1, \ldots, z_n$ are sufficiently close to the zeros of P, then the family of zero-finding methods (4.18) has the order of convergence 4.*

Proof. Let us introduce the abbreviation $\sigma_i = \sum_{j \neq i} \dfrac{W_j}{\zeta_i - z_j}$. Since

$$W_j = (z_j - \zeta_j) \prod_{k \neq j} \frac{z_j - \zeta_k}{z_j - z_k},$$

we have the estimates

$$W_i = \mathcal{O}_M(u_i) = \mathcal{O}_M(u_*), \quad G_{1,i} = \mathcal{O}_M(u_*), \quad G_{2,i} = \mathcal{O}_M(u_*),$$

$$\sigma_i = \mathcal{O}_M(u_*), \quad t_i = \mathcal{O}_M(u_*^2). \tag{4.26}$$

Let z be a complex number such that $|z| < 1$. Then, we have the developments

$$\sqrt{1+z} = 1 + \frac{z}{2} - \frac{z^2}{8} + \cdots \quad \text{and} \quad (1+z)^{-1} = 1 - z + z^2 - z^3 + \cdots, \tag{4.27}$$

where the principal branch is taken in the case of the square root.

Starting from (4.24) and using the developments (4.27), we find

$$\hat{z}_i = z_i - \frac{(\alpha + 1)W_i}{\alpha(1 + G_{1,i}) + (1 + G_{1,i})\sqrt{1 + 2(\alpha + 1)t_i}}$$

$$= z_i - \frac{(\alpha + 1)W_i}{(1 + G_{1,i})(\alpha + 1 + (\alpha + 1)t_i + \mathcal{O}_M(t_i^2))},$$

wherefrom (assuming that $\alpha \neq -1$)

$$\hat{z}_i = z_i - \frac{W_i}{(1 + G_{1,i})(1 + t_i + \mathcal{O}_M(t_i^2))}$$

$$= z_i - \frac{W_i}{1 + G_{1,i}} \left(1 - \frac{W_i G_{2,i}}{(1 + G_{1,i})^2} + \mathcal{O}_M(t_i^2) \right). \tag{4.28}$$

Setting $z := \zeta_i$ in (4.16) (giving $W_i(\zeta_i) = 0$), we obtain $W_i = u_i(1 + \sigma_i)$. Now, (4.28) becomes

$$\hat{z}_i = z_i - \frac{u_i(1 + \sigma_i)}{1 + G_{1,i}} \left(1 - \frac{u_i(1 + \sigma_i)G_{2,i}}{\left(1 + G_{1,i}\right)^2} + \mathcal{O}_M(t_i^2) \right),$$

wherefrom, taking into account the estimates (4.26),

$$\hat{u}_i = \hat{z}_i - \zeta_i = u_i - \frac{u_i(1 + \sigma_i)\left[(1 + G_{1,i})^2 - u_i(1 + \sigma_i)G_{2,i}\right]}{(1 + G_{1,i})^3} + \mathcal{O}_M(u_*^5).$$

After short rearrangement, we find

$$\hat{u}_i = \frac{u_i(X_i + Y_i + Z_i)}{(1 + G_{1,i})^3} + \mathcal{O}_M(u_*^5), \tag{4.29}$$

where

$$
\begin{aligned}
X_i &= G_{1,i} - \sigma_i + u_i G_{2,i}, \\
Y_i &= G_{1,i}\left[2(G_{1,i} - \sigma_i) + G_{1,i}^2 - G_{1,i}\sigma_i\right], \\
Z_i &= u_i G_{2,i}\sigma_i(2 + \sigma_i).
\end{aligned}
$$

Since

$$G_{1,i} - \sigma_i = \sum_{j \neq i} \frac{W_j}{z_i - z_j} - \sum_{j \neq i} \frac{W_j}{\zeta_i - z_j} = -u_i \sum_{j \neq i} \frac{W_j}{(\zeta_i - z_j)(z_i - z_j)}, \tag{4.30}$$

we have

$$G_{1,i} - \sigma_i + u_i G_{2,i} = u_i^2 \sum_{j \neq i} \frac{W_j}{(z_i - z_j)^2(\zeta_i - z_j)}. \tag{4.31}$$

According to (4.26), (4.30), and (4.31), we estimate

$$X_i = u_i^2 \mathcal{O}_M(u_*), \quad Y_i = \mathcal{O}_M(u_*^3), \quad Z_i = u_i \mathcal{O}_M(u_*^2). \tag{4.32}$$

The denominator $(1 + G_{1,i})^3$ tends to 1 when the errors $u_1, \ldots, u_n$ tend to 0. Having in mind this fact and the estimates (4.32), from (4.29), we find $\hat{u}_i = u_i \mathcal{O}_M(u_*^3) = \mathcal{O}_M(u_*^4)$, which completes the proof of the theorem. $\square$

Initial Conditions and Guaranteed Convergence

In the previous analysis, we have proved that the order of convergence of the family (4.18) is 4, assuming that the initial approximations are sufficiently small, neglecting details concerned with the distribution of approximations and their closeness to the corresponding zeros. We apply now Theorem 3.1 and initial conditions of the form (3.4) to state the convergence theorem for the one-parameter family (4.18) of simultaneous methods for finding polynomial zeros (see [117], [132]). The initial condition is computationally verifiable and guarantees the convergence, which is of importance in practice. Before establishing the main result, we give two necessary lemmas.

Lemma 4.1. Let $z_1, \ldots, z_n$ be distinct approximations to the zeros $\zeta_1, \ldots, \zeta_n$ of a polynomial P of degree $n \geq 3$ and let $\hat{z}_1, \ldots, \hat{z}_n$ be new approximations obtained by the iterative formula (4.18). If the inequality

$$w \leq c_n d, \quad c_n \in \left(0, \frac{2}{5(n-1)}\right], \tag{4.33}$$

holds, then for $i, j \in I_n$ we obtain:

(i) $\dfrac{c_n}{(1 - 2q_n)\lambda_n} = 1 - (n-1)c_n \leq |1 + G_{1,i}| \leq 1 + (n-1)c_n$

$$= 2 - \frac{c_n}{(1 - 2q_n)\lambda_n},$$

(ii) $|G_{2,i}| \leq \dfrac{(n-1)w}{d^2}$,

(iii) $|t_i| \leq q_n$,

(iv) $\sqrt{1 + 2(\alpha+1)t_i} \in \left\{ 1; \dfrac{2|\alpha+1|q_n}{1 + h_n} \right\}$,

(v) $|\hat{z}_i - z_i| = |C_i| \leq \dfrac{\lambda_n}{c_n}|W_i| \leq \lambda_n d$,

where

$$\lambda_n = \frac{c_n}{(1 - 2q_n)(1 - (n-1)c_n)}.$$

Proof. According to the definition of the minimal distance d and the inequality (4.33), we have

$$\sum_{j \neq i} \frac{|W_j|}{|z_i - z_j|} \leq \frac{(n-1)w}{d} \leq (n-1)c_n,$$

so that we estimate

$$|1 + G_{1,i}| \geq 1 - \sum_{j \neq i} \frac{|W_j|}{|z_i - z_j|} \geq 1 - (n-1)c_n = \frac{c_n}{(1 - 2q_n)\lambda_n},$$

$$|1 + G_{1,i}| \leq 1 + \sum_{j \neq i} \frac{|W_j|}{|z_i - z_j|} \leq 1 + (n-1)c_n = 2 - \frac{c_n}{(1 - 2q_n)\lambda_n},$$

$$|G_{2,i}| \leq \sum_{j \neq i} \frac{|W_j|}{|z_i - z_j|^2} \leq \frac{(n-1)w}{d^2}.$$

Thus, the assertions (i) and (ii) of Lemma 4.1 are proved.

Using (i), (ii), and (4.33), we prove (iii):

$$|t_i| = \left| \frac{W_i G_{2,i}}{(1 + G_{1,i})^2} \right| \leq \frac{(n-1)w^2}{d^2(1 - (n-1)c_n)^2} \leq \frac{(n-1)c_n^2}{(1 - (n-1)c_n)^2} = q_n.$$

Hence, we conclude that $t_i \in T := \{0; q_n\}$, where T is the disk centered at the origin with the radius q_n. Using the inclusion isotonicity property (1.70) and the formula (1.72) for the square root of a disk (taking the principal value-set centered at $\sqrt{|c|}e^{i\frac{\theta}{2}}$), we find

$$\sqrt{1 + 2(\alpha+1)t_i} \in \sqrt{1 + 2|\alpha+1|T} = \sqrt{\{1; 2|\alpha+1|q_n\}} = \left\{ 1; \frac{2|\alpha+1|q_n}{1 + h_n} \right\},$$

which proves the assertion (iv) of the lemma.

Assume that $\alpha \neq -1$, which means that we deal with the square root in (4.24). By means of (iv) and applying the centered inversion of a disk (1.63), we obtain

$$\frac{\alpha+1}{(1+G_{1,i})(\alpha+\sqrt{1+2(\alpha+1)t_i}\,)} \in \frac{\alpha+1}{(1+G_{1,i})\left(\alpha+\left\{1;\dfrac{2|\alpha+1|q_n}{1+h_n}\right\}\right)}$$

$$= \frac{1}{(1+G_{1,i})\left\{1;\dfrac{2q_n}{1+h_n}\right\}}$$

$$\subseteq \frac{1}{1+G_{1,i}}\left\{1;\frac{2q_n}{1+h_n-2q_n}\right\}.$$

Using the last inclusion, from the iterative formula (4.24), we find

$$\hat{z}_i - z_i = -C_i = -\frac{(\alpha+1)W_i}{(1+G_{1,i})(\alpha+\sqrt{1+2(\alpha+1)t_i}\,)}$$

$$\in -\frac{W_i}{1+G_{1,i}}\left\{1;\frac{2q_n}{1+h_n-2q_n}\right\},$$

where C_i is the iterative correction appearing in (4.24). According to the inequality (1.71), from the last expression, it follows that

$$|\hat{z}_i - z_i| = |C_i| < \left|\frac{W_i}{1+G_{1,i}}\right|\left(1+\frac{2q_n}{1+h_n-2q_n}\right) = \frac{|W_i/(1+G_{1,i})|}{1-2q_n/(1+h_n)}. \quad (4.34)$$

Using the inequalities (4.33), $|1+G_{1,i}| \geq 1-(n-1)c_n$, and $\dfrac{2q_n}{1+h_n} \leq 2q_n$, we start from (4.34) and prove (v)

$$|\hat{z}_i - z_i| = |C_i| \leq \frac{|W_i|}{1-(n-1)c_n}\frac{1}{1-2q_n} = \frac{\lambda_n}{c_n}|W_i| \leq \lambda_n d.$$

In the case $\alpha = -1$, the inequalities (i)–(iii) of Lemma 4.1 are also valid so that we have from (4.25)

$$|\hat{z}_i - z_i| \leq \left|\frac{W_i}{1+G_{1,i}}\right|\frac{1}{1-|t_i|} \leq \frac{\lambda_n}{c_n}|W_i| \leq \lambda_n d,$$

which coincides with (v) of Lemma 4.1 for $\alpha \neq -1$. $\quad\square$

Lemma 4.2. *Let distinct approximations $z_1,\ldots,z_n$ satisfy the conditions (4.33) and let*

$$\delta_n := \left(\frac{(n-1)\lambda_n^2}{1-\lambda_n}+\frac{14\lambda_n q_n}{5c_n}\right)\left(1+\frac{\lambda_n}{1-2\lambda_n}\right)^{n-1} \leq 1-2\lambda_n. \quad (4.35)$$

Then:

(i) $|\widehat{W}_i| \leq \delta_n |W_i|$.
(ii) $\hat{w} \leq c_n \hat{d}$.

Proof. Putting $z = \hat{z}_i$ in (4.15), we obtain

$$P(\hat{z}_i) = \left(\frac{W_i}{\hat{z}_i - z_i} + 1 + \sum_{j \neq i} \frac{W_j}{\hat{z}_i - z_j} \right) \prod_{j=1}^{n} (\hat{z}_i - z_j).$$

After dividing by $\prod_{j \neq i} (\hat{z}_i - \hat{z}_j)$, we find

$$\widehat{W}_i = \frac{P(\hat{z}_i)}{\prod_{j \neq i}(\hat{z}_i - \hat{z}_j)} = (\hat{z}_i - z_i) \left(\frac{W_i}{\hat{z}_i - z_i} + 1 + \sum_{j \neq i} \frac{W_j}{\hat{z}_i - z_j} \right) \prod_{j \neq i} \left(1 + \frac{\hat{z}_j - z_j}{\hat{z}_i - \hat{z}_j} \right).$$

$$(4.36)$$

First, let $\alpha \neq -1$. Using (i) and (iv) of Lemma 4.1 and the inequality $c_n \leq 2/(5(n-1))$, from the iterative formula (4.24), we obtain by circular arithmetic operations (see Sect. 1.3)

$$\frac{W_i}{\hat{z}_i - z_i} = -\frac{(1 + G_{1,i})\left(\alpha + \sqrt{1 + 2(\alpha + 1)t_i} \right)}{\alpha + 1}$$

$$\in -\frac{(1 + G_{1,i})(\alpha + 1)\left\{ 1; \dfrac{2q_n}{1 + h_n} \right\}}{\alpha + 1}$$

$$= \left\{ -1 - G_{1,i}; |1 + G_{1,i}| \frac{2q}{1 + h_n} \right\}$$

$$\subseteq \left\{ -1 - G_{1,i}; 2q_n(1 + (n-1)c_n) \right\},$$

wherefrom

$$\frac{W_i}{\hat{z}_i - z_i} \subseteq \left\{ -1 - G_{1,i}; \frac{14q_n}{5} \right\}.$$

According to this, we derive the following inclusion:

$$\frac{W_i}{\hat{z}_i - z_i} + 1 + \sum_{j \neq i} \frac{W_j}{\hat{z}_i - z_j} \in \left\{ -1 - G_{1,i}; \frac{14q_n}{5} \right\} + 1 + \sum_{j \neq i} \frac{W_j}{\hat{z}_i - z_j}$$

$$= \left\{ -1 - \sum_{j \neq i} \frac{W_j}{z_i - z_j} + 1 + \sum_{j \neq i} \frac{W_j}{\hat{z}_i - z_j}; \frac{14q_n}{5} \right\}$$

$$= \left\{ -(\hat{z}_i - z_i) \sum_{j \neq i} \frac{W_j}{(z_i - z_j)(\hat{z}_i - z_j)}; \frac{14q_n}{5} \right\}.$$

Hence, by (1.71), we estimate

$$\left|\frac{W_i}{\hat{z}_i - z_i} + 1 + \sum_{j \neq i} \frac{W_j}{\hat{z}_i - z_j}\right| \leq |\hat{z}_i - z_i| \sum_{j \neq i} \frac{|W_j|}{|z_i - z_j||\hat{z}_i - z_j|} + \frac{14q_n}{5}. \quad (4.37)$$

By applying (v) of Lemma 4.1, we have

$$|\hat{z}_i - z_j| \geq |z_i - z_j| - |\hat{z}_i - z_i| \geq d - \lambda_n d = (1 - \lambda_n)d, \quad (4.38)$$

$$|\hat{z}_i - \hat{z}_j| \geq |z_i - z_j| - |\hat{z}_i - z_i| - |\hat{z}_j - z_j| \geq d - 2\lambda_n d = (1 - 2\lambda_n)d. \quad (4.39)$$

From (4.39) and taking into account the definition of the minimal distance, we find

$$\hat{d} \geq (1 - 2\lambda_n)d \quad \text{or} \quad d \leq \frac{\hat{d}}{1 - 2\lambda_n}. \quad (4.40)$$

Using the estimates (4.33), (4.38), and (v) of Lemma 4.1, from (4.37), we obtain

$$\left|\frac{W_i}{\hat{z}_i - z_i} + 1 + \sum_{j \neq i} \frac{W_j}{\hat{z}_i - z_j}\right| \leq \lambda_n d \frac{(n-1)w}{(1 - \lambda_n)d^2} + \frac{14q_n}{5} \leq \frac{(n-1)c_n\lambda_n}{1 - \lambda_n} + \frac{14q_n}{5}. \quad (4.41)$$

By (v) of Lemma 4.1 and (4.39), there follows

$$\left|\prod_{j \neq i}\left(1 + \frac{\hat{z}_j - z_j}{\hat{z}_i - \hat{z}_j}\right)\right| \leq \prod_{j \neq i}\left(1 + \frac{|\hat{z}_j - z_j|}{|\hat{z}_i - \hat{z}_j|}\right) < \left(1 + \frac{\lambda_n d}{(1 - 2\lambda_n)d}\right)^{n-1}$$

$$= \left(1 + \frac{\lambda_n}{1 - 2\lambda_n}\right)^{n-1}. \quad (4.42)$$

Taking into account (v) of Lemma 4.1, (4.41), and (4.42), from (4.36), we obtain

$$|\widehat{W}_i| \leq |\hat{z}_i - z_i|\left|\frac{W_i}{\hat{z}_i - z_i} + 1 + \sum_{j \neq i} \frac{W_j}{\hat{z}_i - z_j}\right|\left|\prod_{j \neq i}\left(1 + \frac{\hat{z}_j - z_j}{\hat{z}_i - \hat{z}_j}\right)\right|$$

$$< \frac{\lambda_n}{c_n}|W_i|\left(\frac{(n-1)c_n\lambda_n}{1 - \lambda_n} + \frac{14q_n}{5}\right)\left(1 + \frac{\lambda_n}{1 - 2\lambda_n}\right)^{n-1},$$

wherefrom

$$|\widehat{W}_i| \leq \delta_n|W_i|, \quad (4.43)$$

and the assertion (i) of Lemma 4.2 is proved. Using this inequality and the inequalities (4.33) and (4.40), we prove the assertion (ii)

$$\widehat{w} \leq \delta_n w \leq \delta_n c_n d \leq \frac{\delta_n c_n}{1 - 2\lambda_n}\hat{d} \leq c_n\hat{d},$$

the last inequality being valid due to $\delta_n \leq 1 - 2\lambda_n$.

Let us consider now the case $\alpha = -1$. From (4.25), we obtain

$$\frac{W_i}{\hat{z}_i - z_i} = -1 - G_{1,i} - t_i(1 + G_{1,i}) = -1 - \sum_{j \neq i} \frac{W_j}{z_i - z_j} - \frac{W_i G_{2,i}}{1 + G_{1,i}},$$

so that

$$\frac{W_i}{\hat{z}_i - z_i} + 1 + \sum_{j \neq i} \frac{W_j}{\hat{z}_i - z_j} = -(\hat{z}_i - z_i) \sum_{j \neq i} \frac{W_j}{(z_i - z_j)(\hat{z}_i - z_j)} - \frac{W_i G_{2,i}}{1 + G_{1,i}}.$$

Using (v) of Lemma 4.1, (4.38), and the inequality

$$\left| \frac{W_i G_{2,i}}{1 + G_{1,i}} \right| \leq \frac{(n-1)w^2}{(1 - (n-1)c_n)d^2} \leq \frac{(n-1)c_n^2}{1 - (n-1)c_n},$$

which follows according to (i)–(iii) of Lemma 4.1 and (4.33), we obtain

$$\left| \frac{W_i}{\hat{z}_i - z_i} + 1 + \sum_{j \neq i} \frac{W_j}{\hat{z}_i - z_j} \right| \leq \frac{(n-1)c_n \lambda_n}{1 - \lambda_n} + \frac{(n-1)c_n^2}{1 - (n-1)c_n}$$

$$\leq \frac{(n-1)c_n \lambda_n}{1 - \lambda_n} + \frac{14 q_n}{5}.$$

The last inequality, together with (4.42), yields the inequality of the form (4.43) with δ_n given by (4.35).

The inequality $\hat{w} \leq c_n \hat{d}$ for $\alpha = -1$ is proved in a similar way. Therefore, as in the case $\alpha \neq -1$, we obtain the inequalities (i) and (ii) of Lemma 4.2. $\qquad \square$

Theorem 4.2. Let $\alpha \in A_n(\alpha)$, $n \geq 3$ and let the condition (4.33) be valid. In addition, let

$$\beta_n := \delta_n \left(\frac{2\lambda_n}{c_n} - \frac{1}{1 - 2q_n} + \frac{14\lambda_n q_n}{5c_n} \right) < 1 \qquad (4.44)$$

and

$$g(\beta_n) < \frac{1}{2\lambda_n}. \qquad (4.45)$$

Then, the one-parameter family (4.18) is convergent.

Proof. It is sufficient to prove that Theorem 3.1 holds true under the conditions (4.33), (4.44), and (4.45).

According to (ii) of Lemma 4.2, we have the implication

$$w^{(0)} \leq c_n d^{(0)} \implies w^{(1)} \leq c_n d^{(1)}.$$

Using the same argumentation, under the given conditions, we derive the implication

$$w^{(m)} \leq c_n d^{(m)} \quad \Longrightarrow \quad w^{(m+1)} \leq c_n d^{(m+1)}.$$

Hence, we prove by induction that (4.33) implies

$$w^{(m)} \leq c_n d^{(m)}$$

for each $m = 1, 2, \ldots$, which means that all assertions of Lemmas 4.1 and 4.2 hold for each $m = 1, 2, \ldots$. In particular, the inequalities

$$\left| W_i^{(m+1)} \right| \leq \delta_n \left| W_i^{(m)} \right| \quad (i \in I_n, \ m = 0, 1, \ldots) \tag{4.46}$$

and

$$\left| C_i^{(m)} \right| = \left| z_i^{(m+1)} - z_i^{(m)} \right| \leq \frac{\lambda_n}{c_n} \left| W_i^{(m)} \right| \quad (i \in I_n, \ m = 0, 1, \ldots) \tag{4.47}$$

are valid.

From (4.24), we see that the iterative correction $C_i^{(m)}$ is given by

$$C_i^{(m)} = \frac{(\alpha + 1) W_i^{(m)}}{\left(1 + G_{1,i}^{(m)}\right) \left(\alpha + \sqrt{1 + 2(\alpha + 1)t_i^{(m)}}\right)}. \tag{4.48}$$

Omitting the iteration index, from (4.47), we obtain by (4.46) and (4.48)

$$|\widehat{C}_i| \leq \frac{\lambda_n}{c_n} |\widehat{W}_i| \leq \frac{\lambda_n \delta_n}{c_n} |C_i| |y_i|, \tag{4.49}$$

where

$$y_i = \frac{\left(1 + G_{1,i}\right) \left(\alpha + \sqrt{1 + 2(\alpha + 1)t_i}\right)}{\alpha + 1}. \tag{4.50}$$

According to (4.33) and (i) of Lemma 4.1, we have

$$|1 + G_{1,i}| \leq 1 + (n - 1)c_n \leq 1 + \frac{2(n - 1)}{5(n - 1)} = \frac{7}{5}.$$

By this bound and (iii) of Lemma 4.1, from (4.50), we find

$$y_i \in (1 + G_{1,i}) \frac{\alpha + \left\{1; \dfrac{2|\alpha + 1|q_n}{1 + h_n}\right\}}{\alpha + 1} = (1 + G_{1,i}) \left\{1; \frac{2q_n}{1 + h_n}\right\}$$

$$\subseteq \left\{1 + G_{1,i}; 2q_n|1 + G_{1,i}|\right\} \subseteq \left\{1 + G_{1,i}; \frac{14q_n}{5}\right\},$$

wherefrom we find the upper bound of $|y_i|$:

$$|y_i| \leq |1 + G_{1,i}| + \frac{14q_n}{5} \leq 2 - \frac{c_n}{(1 - 2q_n)\lambda_n} + \frac{14q_n}{5}.$$

Using this bound, from (4.49), we obtain

$$|\widehat{C}_i| \leq \frac{\lambda_n \delta_n}{c_n} \left(2 - \frac{c_n}{(1 - 2q_n)\lambda_n} + \frac{14q_n}{5} \right) |C_i| = \beta_n |C_i|,$$

where β_n is given by (4.44).

In the case $\alpha = -1$, letting $\alpha \to -1$ in (4.50), we obtain

$$y_i = (1 + G_{1,i})(1 + t_i).$$

According to (i)–(iii) of Lemma 4.1, there follows

$$|y_i| \leq |1 + G_{1,i}|(1 + |t_i|) \leq \left(2 - \frac{c_n}{(1 - 2q_n)\lambda_n} \right)(1 + q_n)$$

$$< 2 - \frac{c_n}{(1 - 2q_n)\lambda_n} + \frac{14q_n}{5}.$$

Therefore, in both cases, $\alpha \neq -1$ and $\alpha = -1$, we have proved the inequality

$$\left| C_i^{(m+1)} \right| \leq \beta_n \left| C_i^{(m)} \right| \quad (i \in \boldsymbol{I}_n, \ m = 0, 1, \ldots),$$

which completes the proof of the assertion (i) of Theorem 3.1 applied to the iterative method (4.18) under the conditions of Theorem 4.2.

To prove (ii) of Theorem 3.1, we use the bound (v) from Lemma 4.1 and find $|C_i^{(0)}| \leq \lambda_n |W_i^{(0)}|/c_n$. According to this, (4.33), and (4.45), one obtains

$$|z_i^{(0)} - z_j^{(0)}| \geq d^{(0)} \geq \frac{w^{(0)}}{c_n} \geq \frac{1}{2\lambda_n} \left(|C_i^{(0)}| + |C_j^{(0)}| \right) > g(\beta_n) \left(|C_i^{(0)}| + |C_j^{(0)}| \right).$$

Finally, we prove that the family of iterative methods (4.24) is well defined in each iteration. From (4.24), we observe that

$$C_i = \frac{P(z_i)}{F_i(z_1, \ldots, z_n)},$$

where

$$F_i(z_1, \ldots, z_n) = \left(\frac{1 + G_{1,i}}{1 + \alpha} \right) \left(\alpha + \sqrt{1 + 2(\alpha + 1)t_i} \right) \prod_{\substack{j=1 \\ j \neq i}}^{n} (z_i - z_j).$$

Let $\alpha \neq -1$. From (iii) of Lemma 4.1, we find

$$\alpha + \sqrt{1 + 2(\alpha + 1)t_i} \in \left\{ \alpha + 1; \frac{2|\alpha + 1|q_n}{1 + h_n} \right\} \subseteq (\alpha + 1)\{1; 2q_n\},$$

where $\{1; 2q_n\}$ is the disk centered at 1 with the radius $2q_n$. Since $|z| \geq 1 - 2q_n$ for any $z \in \{1; 2q_n\}$, it follows

$$\left| \alpha + \sqrt{1 + 2(\alpha + 1)t_i} \right| \geq |\alpha + 1|(1 - 2q_n). \tag{4.51}$$

Taking the upper bound $c_n \leq 2/(5(n - 1))$ (see Lemma 4.1), we estimate

$$q_n \leq \frac{4}{9(n - 1)} \leq \frac{2}{9}. \tag{4.52}$$

According to (4.51), (4.52), and the bound

$$|1 + G_{1,i}| \geq 1 - (n - 1)c_n \geq 1 - (n - 1)\frac{2}{5(n - 1)} = \frac{3}{5} \tag{4.53}$$

(see (i) of Lemma 4.1), from (4.51), we find

$$\left| \left(\frac{1 + G_{1,i}}{1 + \alpha} \right) \left(\alpha + \sqrt{1 + 2(\alpha + 1)t_i} \right) \right| \geq |1 + G_{1,i}|(1 - 2q_n) \geq \frac{3}{5}\left(1 - \frac{4}{9}\right) = \frac{1}{3} > 0.$$

In addition, since $|z_i - z_j| \geq d > 0$, we find that $\prod_{j \neq i}(z_i - z_j) \neq 0$, and thus, $F_i(z_1, \ldots, z_n) \neq 0$.

In a similar way, we derive the proof in the case $\alpha = -1$. Namely, then

$$F_i(z_1, \ldots, z_n) = (1 + G_{1,i})(1 + t_i) \prod_{j \neq i}(z_i - z_j),$$

so that by (4.53) and (ii) of Lemma 4.1 we find

$$|F_i(z_1, \ldots, z_n)| = |1 + G_{1,i}||1 + t_i| \prod_{j \neq i}|z_i - z_j| \geq \left(1 - (n - 1)c_n\right)(1 - q_n)d^{n-1}$$

$$\geq \frac{3}{5} \cdot \frac{7}{9}d^{n-1} = \frac{7}{15}d^{n-1} > 0. \quad \square$$

Theorem 4.3. *Let* $z_1^{(0)}, \ldots, z_n^{(0)}$ *be distinct initial approximations satisfying the initial condition*

$$w^{(0)} \leq \frac{d^{(0)}}{2.7n + 0.75}. \tag{4.54}$$

Then, the family of simultaneous methods (4.18) is convergent.

Proof. Having in mind the assertion of Theorem 4.2, it is sufficient to prove that the specific value of i-factor $c_n = 1/(2.7n + 0.75)$, appearing in (4.54), satisfies the conditions (4.33), (4.35), (4.44), and (4.45).

First, we directly verify that $c_n = 1/(2.7n + 0.75) \in \left(0, 2/(5(n - 1))\right]$. Furthermore, the sequence $\{\delta_n\}$, given by (4.35), is monotonically decreasing for $n \geq 4$ and $\delta_n \leq \delta_4 < 0.362$ $(n \geq 4)$ holds. In addition, $\delta_3 < 0.365$ so that $\delta_n < 1$. Besides, $1 - 2\lambda_n > 0.68$ $(n \geq 3)$ and, therefore, $\delta_n < 0.365 <$

$1 - 2\lambda_n$, which means that (4.35) holds. The sequence $\{\beta_n\}$ is monotonically decreasing for $n \geq 4$ and

$$1 > \beta_3 = 0.66352\ldots, \quad 0.731 > \beta_4 \geq \beta_n \ (n \geq 4),$$

i.e., the condition (4.44) is fulfilled.

Finally, the sequence $\{g(\beta_n) - 1/2\lambda_n\}$ is monotonically decreasing for $n \geq 4$ and

$$g(\beta_n) - \frac{1}{2\lambda_n} \leq g(\beta_4) - \frac{1}{2\lambda_4} = -0.211826\ldots < 0$$

is valid. In particular, $g(\beta_3) - \frac{1}{2\lambda_3} = -0.161\ldots < 0$. Therefore, the inequality $g(\beta_n) < \frac{1}{2\lambda_n}$ holds for all $n \geq 3$, and thus, the condition (4.47) is also satisfied. This completes the proof of the convergence theorem. $\square$

4.3 Family of Simultaneous Methods with Derivatives

Let us consider Hansen–Patrick's family (4.1) and let $f \equiv P$ be a monic polynomial of order n with (real or complex) simple zeros. Obviously, the zeros of P coincide with the zeros of the rational Weierstrass' function $W_i(z)$ given by (1.17). In the subsequent discussion, we will use the following abbreviations:

$$\delta_{k,i} = \frac{P^{(k)}(z_i)}{P(z_i)}, \quad S_{k,i} = \sum_{\substack{j=1 \\ j \neq i}}^{n} \frac{1}{(z_i - z_j)^k} \quad (k = 1, 2).$$

Applying Hansen–Patrick's formula (4.1) to the function $W_i(z)$ (see (1.17)) and using (1.20), the following one-parameter family of iterative methods for the simultaneous approximation of all simple zeros of a polynomial P has been derived by M. Petković, Sakurai, and Rančić [141]:

$$\hat{z}_i = z_i - \frac{\alpha + 1}{\alpha(\delta_{1,i} - S_{1,i}) + \left[(\alpha + 1)(\delta_{1,i}^2 - \delta_{2,i} - S_{2,i}) - \alpha(\delta_{1,i} - S_{1,i})^2\right]_*^{1/2}}$$

$$(i \in \boldsymbol{I}_n). \ (4.55)$$

It is assumed that two values of the (complex) square root are taken in (4.55). We have to choose a "proper" sign in front of the square root in such a way that a smaller step $|\hat{z}_i - z_i|$ is taken. The symbol $*$ in (4.55) and subsequent iterative formulae points to the proper sign. A criterion for the selection of the proper value of the square root, which has a practical importance, can be stated according to the result of Henrici [57, p. 532], see Remark 4.2. If the approximations are sufficiently close to the zeros, it turns out that one has to choose the principal branch of the square root in (4.55).

We present some special cases of the iterative formula (4.55):

$\alpha = 0$, *the Ostrowski-like method:*

$$\hat{z}_i = z_i - \frac{1}{\left[\delta_{1,i}^2 - \delta_{2,i} - S_{2,i}\right]_*^{1/2}} \quad (i \in \boldsymbol{I}_n).$$

This method was derived by Gargantini as a special case of the square root interval method in circular complex arithmetic [47].

$\alpha = 1$, *the Euler-like method:*

$$\hat{z}_i = z_i - \frac{2}{\delta_{1,i} - S_{1,i} + \left[2(\delta_{1,i}^2 - \delta_{2,i} - S_{2,i}) - (\delta_{1,i} - S_{1,i})^2\right]_*^{1/2}} \quad (i \in \boldsymbol{I}_n).$$

$\alpha = 1/(n-1)$, *the Laguerre-like method:*

$$\hat{z}_i = z_i - \frac{n}{\delta_{1,i} - S_{1,i} + \left[n(n-1)(\delta_{1,i}^2 - \delta_{2,i} - S_{2,i}) - (n-1)(\delta_{1,i} - S_{1,i})^2\right]_*^{1/2}}$$
$$(i \in \boldsymbol{I}_n).$$

$\alpha = -1$, *the Halley-like method:*

$$\hat{z}_i = z_i - \frac{2(S_{1,i} - \delta_{1,i})}{\delta_{2,i} - 2\delta_{1,i}^2 + 2S_{1,i}\delta_{1,i} + S_{2,i} - S_{1,i}^2} \quad (i \in \boldsymbol{I}_n). \tag{4.56}$$

The Halley-like method (4.56) is obtained from (4.55) for $\alpha \to -1$ applying a limiting operation. This method was also derived by Sakurai, Torii, and Sugiura [157] using a different approach.

$\alpha \to \infty$, *the Newton-like method* or *Ehrlich–Aberth's method* ([1], [33]) of the third order:

$$\hat{z}_i = z_i - \frac{1}{\delta_{1,i} - S_{1,i}} \quad (i \in \boldsymbol{I}_n). \tag{4.57}$$

Remark 4.6. Let $f \equiv P$ be a monic polynomial of order n with simple zeros and let c_j be either $c_j = z_j - N_j$ (the Newton's approximation) or $c_j = z_j - H_j$ (the Halley's approximation) (see (1.18)). Applying Hansen–Patrick's formula (4.1) to the modified Weierstrass' function given by (1.19), the following one-parameter family of iterative methods for the simultaneous approximation of all simple zeros of a polynomial P is derived:

$$\hat{z}_i = z_i - \frac{\alpha + 1}{\alpha(\delta_{1,i} - S_{1,i}^*) + \left[(\alpha + 1)(\delta_{1,i}^2 - \delta_{2,i} - S_{2,i}^*) - \alpha\left[\delta_{1,i} - \left(S_{1,i}^*\right)^2\right]\right]_*^{1/2}}$$

$$(i \in I_n),$$

where $S_{k,i}^* = \sum_{j \neq i}(z_i - c_j)^{-k}$ $(k = 1, 2)$. This is a new family with a high order of convergence. In the special case $\alpha = 0$, it reduces to the improved iterative methods (1.27) and (1.28) of Ostrowski's type. If we take $c_j = z_j$, then we obtain the basic simultaneous method of the fourth order (4.55), considered in [141].

In this section, we investigate iterative methods from the family (4.55), which have the order of convergence 4, see Theorem 4.4. For this reason, we will assume that the parameter α is not too large in magnitude, see Remarks 4.5 and 4.8 and the iterative formula (4.57).

Convergence Analysis

Now, we determine the convergence order of the family of simultaneous methods (4.55). In addition, according to the discussion given in Remarks 4.5 and 4.8, we will assume that the parameter α is not too large in magnitude.

Theorem 4.4. *If initial approximations* $z_1^{(0)}, \ldots, z_n^{(0)}$ *are sufficiently close to the zeros* $\zeta_1, \ldots, \zeta_n$ *of the polynomial P, then the family of simultaneous iterative methods* (4.55) *has the order of convergence equal to 4.*

Proof. Let us introduce the errors $u_i = z_i - \zeta_i$, $\hat{u}_i = \hat{z}_i - \zeta_i$ $(i \in I_n)$ and the abbreviations

$$A_i = \sum_{\substack{j=1 \\ j \neq i}}^n \frac{u_j}{(z_i - \zeta_j)(z_i - z_j)}, \qquad B_i = \sum_{\substack{j=1 \\ j \neq i}}^n \frac{(2z_i - z_j - \zeta_j)u_j}{(z_i - \zeta_j)^2(z_i - z_j)^2}.$$

In our proof, we will use the identities

$$\delta_{1,i} = \sum_{j=1}^n \frac{1}{z_i - \zeta_j}, \qquad \delta_{1,i}^2 - \delta_{2,i} = \sum_{j=1}^n \frac{1}{(z_i - \zeta_j)^2}.$$

Hence, after some elementary calculations, we obtain

$$\delta_{1,i} - S_{1,i} = \frac{1}{u_i}\left(1 - A_i u_i\right), \qquad \delta_{1,i}^2 - \delta_{2,i} - S_{2,i} = \frac{1}{u_i^2}\left(1 - B_i u_i^2\right).$$

Using the last two relations, from the iterative formula (4.55), we find

$$\hat{u}_i = \hat{z}_i - \zeta_i = u_i - \frac{(\alpha+1)u_i}{\alpha(1-A_iu_i) + \left[(\alpha+1)(1-B_iu_i^2) - \alpha(1-A_iu_i)^2\right]_*^{1/2}}$$

$$= u_i - \frac{(\alpha+1)u_i}{\alpha(1-A_iu_i) + \left[1+V_iu_i\right]_*^{1/2}},$$

where

$$V_i = 2\alpha A_i - (\alpha+1)B_iu_i - \alpha A_i^2 u_i.$$

Assuming that $|u_i|$ is sufficiently small and having in mind the principal branch of the square root, we have

$$\sqrt{1+V_iu_i} = 1 + \frac{V_iu_i}{2} + \mathcal{O}_M\left(V_i^2u_i^2\right)$$

(see (4.27)). Now, we obtain

$$\hat{u}_i = u_i - \frac{(\alpha+1)u_i}{\alpha(1-A_iu_i) + 1 + \dfrac{V_iu_i}{2} + \mathcal{O}_M\left(V_i^2u_i^2\right)}$$

$$= u_i - \frac{2(\alpha+1)u_i}{2(\alpha+1) - (\alpha+1)B_iu_i^2 - \alpha A_i^2 u_i^2 + \mathcal{O}_M\left(V_i^2u_i^2\right)},$$

or, after a short rearrangement,

$$\hat{u}_i = \frac{u_i^3\left((\alpha+1)B_i + \alpha A_i^2\right) + \mathcal{O}_M\left(V_i^2u_i^3\right)}{(\alpha+1)B_iu_i^2 + \alpha A_i^2 u_i^2 - 2(\alpha+1) + \mathcal{O}_M\left(V_i^2u_i^2\right)}. \tag{4.58}$$

Let $u = \max\limits_{1\le j\le n} |u_j|$. We estimate $|A_i| = \mathcal{O}(u)$, $|B_i| = \mathcal{O}(u)$, and $|V_i| = \mathcal{O}(u)$, so that from (4.58) there follows

$$|\hat{u}_i| = |u_i|^3\mathcal{O}(u).$$

If we adopt that absolute values of all errors u_j $(j = 1,\dots,n)$ are of the same order, say $|u_j| = \mathcal{O}(|u_i|)$ for any pair $i,j \in I_n$, we will have

$$|\hat{u}| = \mathcal{O}\left(|u|^4\right),$$

which proves the assertion. □

Remark 4.7. Recently, Huang and Zheng [62] have used the approximation

$$\frac{P''(z_i)}{P'(z_i)} \approx 2\sum_{\substack{j=1 \\ j\neq i}}^{n} \frac{1}{z_i - z_j}$$

in (4.1) to derive a new family of simultaneous methods of the form

$$\hat{z}_i = z_i - \frac{(\alpha+1)/\delta_{1,i}}{\alpha + \left[1 - \frac{2(\alpha+1)}{\delta_{1,i}} \sum_{\substack{j=1 \\ j \neq i}}^{n} \frac{1}{z_i - z_j}\right]^{1/2}} \quad (i \in \boldsymbol{I}_n). \tag{4.59}$$

This iterative formula is simpler than (4.55) but it possesses only a cubic convergence. Besides, a theoretical analysis as well as numerous numerical experiments have shown that the domain of convergence of the family (4.59) is narrower compared with the domain of the family (4.55).

Methods for Multiple Zeros

Let us consider now the case when a polynomial P has multiple zeros $\zeta_1, \ldots, \zeta_\nu$ ($\nu \leq n$) of the known multiplicities $\mu_1, \ldots, \mu_\nu$, respectively. Then, the zeros of P coincide with the zeros of the rational Weierstrass-like function given by (1.40) (with $c_j = z_j$). We note that efficient procedures for finding the order of multiplicity can be found in [78], [79], and [93].

Let $\delta_{q,i}$ ($q = 1, 2$) be defined as above and let

$$\widetilde{S}_{q,i} = \sum_{\substack{j=1 \\ j \neq i}}^{\nu} \frac{\mu_j}{(z_i - z_j)^q} \quad (q = 1, 2). \tag{4.60}$$

Starting from (1.40) and applying logarithmic derivative, using (4.60), we find

$$\frac{\left(W_i^*(z)\right)'}{W_i^*(z)}\bigg|_{z=z_i} = \delta_{1,i} - \widetilde{S}_{1,i}, \tag{4.61}$$

$$\frac{\left(W_i^*(z)\right)''}{\left(W_i^*(z)\right)'}\bigg|_{z=z_i} = \delta_{1,i} - \widetilde{S}_{1,i} + \frac{\delta_{2,i} - \delta_{1,i}^2 + \widetilde{S}_{2,i}}{\delta_{1,i} - \widetilde{S}_{1,i}}. \tag{4.62}$$

Let us rewrite the iterative formula (4.13) (substituting f with P) to the form

$$\hat{z} = z - \frac{\mu(\alpha+1)}{\alpha\frac{P'(z)}{P(z)} + \left[(\mu(\alpha+1) - \alpha)\left(\frac{P'(z)}{P(z)}\right)^2 - \mu(\alpha+1)\frac{P''(z)}{P'(z)} \cdot \frac{P(z)}{P'(z)}\right]^{1/2}_*}.$$

Similarly as in the case of simple zeros, let us substitute P'/P with $(W_i^*)'/W_i^*$ (formula (4.61)), P''/P' with $(W_i^*)''/(W_i^*)'$ (formula (4.62)), and, in addition, μ with μ_i. In this way, we obtain a one-parameter family for the simultaneous determination of multiple zeros (with known multiplicities) of the polynomial P

$$\hat{z}_i = z_i - \frac{\mu_i(\alpha+1)}{\alpha(\delta_{1,i} - \widetilde{S}_{1,i}) + \left[\mu_i(\alpha+1)(\delta_{1,i}^2 - \delta_{2,i} - \widetilde{S}_{2,i}) - \alpha(\delta_{1,i} - \widetilde{S}_{1,i})^2\right]_*^{1/2}}, \tag{4.63}$$

where $i \in \boldsymbol{I}_\nu := \{1,\ldots,\nu\}$. We note that another iterative formula for multiple zeros with similar structure was proposed in [124]. For specific values of the parameter α, we obtain some special cases of the iterative formula (4.63). For example, for $\alpha = 0$ (Ostrowski's case), the iterative formula (4.63) becomes

$$\hat{z}_i = z_i - \frac{\sqrt{\mu_i}}{\left[\delta_{1,i}^2 - \delta_{2,i} - \widetilde{S}_{2,i}\right]_*^{1/2}} \quad (i \in \boldsymbol{I}_\nu).$$

This method arises from the square root iterative interval method proposed by Gargantini in [49] for multiple zeros. Furthermore, in a limiting procedure when $\alpha \to -1$, from (4.63), we obtain the Halley-like method

$$\hat{z}_i = z_i - \frac{2\mu_i(\delta_{1,i} - \widetilde{S}_{1,i})}{(\delta_{1,i} - \widetilde{S}_{1,i})^2 - \mu_i(\delta_{2,i} - \delta_{1,i}^2 + \widetilde{S}_{2,i})} \quad (i \in \boldsymbol{I}_\nu),$$

previously derived in [144] using a different approach.

In a similar way, we can obtain the Euler-like ($\alpha = 1$) and Laguerre-like method ($\alpha = \mu_i/(n - \mu_i)$) for finding multiple zeros. If $\alpha \to \infty$, performing a limiting operation in (4.63), we obtain

$$\hat{z}_i = z_i - \frac{\mu_i}{\delta_{1,i} - \widetilde{S}_{1,i}} = z_i - \frac{\mu_i}{\dfrac{P'(z_i)}{P(z_i)} - \displaystyle\sum_{\substack{j=1 \\ j \neq i}}^{\nu} \dfrac{\mu_j}{z_i - z_j}} \quad (i \in \boldsymbol{I}_\nu),$$

the well-known third-order method for multiple zeros which can be obtained from Gargantini's interval method for the inclusion of multiple zeros [48]. This iterative method can be regarded as Ehrlich–Aberth's version for multiple zeros.

Using a similar analysis as in the proof of Theorem 4.1, we can prove the following assertion.

Theorem 4.5. *If initial approximations $z_1^{(0)},\ldots,z_\nu^{(0)}$ are sufficiently close to the zeros $\zeta_1,\ldots,\zeta_\nu$ of the polynomial P, then the family of simultaneous iterative methods (4.63) has the fourth order of convergence.*

Initial Conditions and Guaranteed Convergence

We have proved above that the family (4.55) possesses the fourth order of convergence, assuming that initial approximations are close enough to the sought zeros. Similarly as in Sect. 4.2, now we give a more precise convergence analysis which includes computationally verifiable initial conditions (see [123]).

In the sequel, we will assume that the following inequality

$$w < c_n d, \quad c_n = \frac{1}{3n+2},$$
(4.64)

is valid. Since $w < 1/(3n+2) < d/(2n)$, the assertions of Corollary 1.1 hold. Let us introduce

$$L_i = \frac{\alpha+1}{\alpha + \left[1 - (\alpha+1)t_i\right]_*^{1/2}}, \quad t_i = 1 + \left(\delta_{2,i} - \delta_{1,i}^2 + S_{2,i}\right)\left(\delta_{1,i} - S_{1,i}\right)^{-2}.$$

Then, the iterative formula (4.55) can be rewritten in the form

$$
\begin{aligned}
\hat{z}_i &= z_i - \frac{(\alpha+1)(\delta_{1,i} - S_{1,i})^{-1}}{\alpha + \left[1 - (\alpha+1)\left(1 + \left(\delta_{2,i} - \delta_{1,i}^2 + S_{2,i}\right)\left(\delta_{1,i} - S_{1,i}\right)^{-2}\right)\right]_*^{1/2}} \\
&= z_i - \frac{(\alpha+1)(\delta_{1,i} - S_{1,i})^{-1}}{\alpha + \left[1 - (\alpha+1)t_i\right]_*^{1/2}},
\end{aligned}
$$

i.e.,

$$\hat{z}_i = z_i - L_i(\delta_{1,i} - S_{1,i})^{-1}.$$
(4.65)

To ensure only positive values under the square root of some quantities, the parameter α must belong to the disk

$$K_\alpha(n) := \{-1; 1/q_n\}, \quad \text{where } q_n = \frac{1}{2n+4} \leq \frac{1}{10}$$
(4.66)

(see [123]).

Remark 4.8. According to (4.66), the values of the parameter α must belong to the disk $K_\alpha(n) = \{-1, 1/q_n\} \subseteq \{-1; 10\}$. However, the above restriction on α should not be regarded as a disadvantage. It can be shown that the use of large values of $|\alpha|$ generates the methods whose convergence rate decreases and approaches 3. In practice, in such situations, the iterative method (4.55) behaves as the aforementioned cubically convergent Ehrlich–Aberth's method (4.57).

Lemma 4.3. *Let (4.64) hold and let $\alpha \in K_\alpha(n) = \{-1; 1/q_n\}$. Then for all $i = 1, \ldots, n$ we have:*

(i) $L_i \in \{1; q_n\}^{-1} \subset \left\{1; \dfrac{q_n}{1 - q_n}\right\}.$

(ii) $|(\delta_{1,i} - S_{1,i})^{-1}| < \frac{3}{2}|W_i|.$

(iii) $|\hat{z}_i - z_i| = |C_i| < \frac{5}{3}|W_i| < \lambda_n d, \quad \text{where } \lambda_n = \dfrac{5}{3(3n+2)}.$

(iv) $|\widehat{W}_i| < \frac{1}{4}|W_i|.$

(v) $\widehat{w} < c_n \widehat{d}, \quad c_n = 1/(3n+2)$.
(vi) $\widehat{d} > (1 - 2\lambda_n)d$.

The proofs of these assertions can be found in [123].

Theorem 4.6. *Let $n \geq 3$ and $\alpha \in K_\alpha(n)$. Then, the one-parameter family of iterative methods (4.55) is convergent under the condition*

$$w^{(0)} < \frac{d^{(0)}}{3n+2}. \tag{4.67}$$

Proof. The following implication arises from Lemma 4.3 (assertion (v)):

$$w < c_n d \implies \widehat{w} < c_n \widehat{d}, \quad c_n = \frac{1}{3n+2}.$$

By induction, we can prove that the condition (4.67) implies the inequality $w^{(m)} < c_n d^{(m)}$ for each $m = 1, 2, \ldots$. Therefore, all the assertions of Lemma 4.3 hold for each $m = 1, 2, \ldots$ if the initial condition (4.67) is valid. In particular, the following inequalities

$$|W_i^{(m+1)}| < \frac{1}{4}|W_i^{(m)}| \tag{4.68}$$

and

$$|C_i^{(m)}| = |z_i^{(m+1)} - z_i^{(m)}| < \frac{5}{3}|W_i^{(m)}| \tag{4.69}$$

hold for $i \in I_n$ and $m = 0, 1, \ldots$.

From the iterative formula (4.65) for $\alpha \neq -1$, we see that the corrections $C_i^{(m)}$ are expressed by

$$C_i^{(m)} = L_i^{(m)} \left(\delta_{1,i}^{(m)} - S_{1,i}^{(m)} \right)^{-1}, \tag{4.70}$$

where the abbreviations $L_i^{(m)}$, $\delta_{1,i}^{(m)}$, and $S_{1,i}^{(m)}$ are related to the mth iterative step.

To prove that the iterative process (4.55) is well defined in each iteration, it is sufficient to show that the function $F_i(z_1, \ldots, z_n) = P(z_i)/C_i$ cannot take the value 0. From (4.70), we have

$$F_i(z_1, \ldots, z_n) = \frac{P(z_i)(\delta_{1,i} - S_{1,i})}{L_i} = \frac{W_i(\delta_{1,i} - S_{1,i})}{L_i} \prod_{j \neq i}(z_i - z_j).$$

Starting from Lemma 4.3(i), we find by (1.63) and (1.71)

$$\frac{1}{L_i} \in \{1; q\} \implies \frac{1}{|L_i|} \geq 1 - q_n = \frac{9}{10}.$$

According to Lemma 4.3(ii), we have

$$|W_i(\delta_{1,i} - S_{1,i}| > |W_i| \cdot \frac{2}{3|W_i|} = \frac{2}{3}.$$

Finally, using the definition of the minimal distance, one obtains

$$\left|\prod_{j \neq i}(z_i - z_j)\right| \geq d^{n-1} > 0.$$

Using the last three inequalities, we get

$$|F_i(z_1, \ldots, z_n)| = \frac{1}{|L_i|}|W_i(\delta_{1,i} - S_{1,i})|\left|\prod_{j \neq i}(z_i - z_j)\right| > \frac{9}{10} \cdot \frac{2}{3} \cdot d^{n-1} > 0.$$

Now, we prove that the sequences $\{|C_i^{(m)}|\}$ ($i \in \boldsymbol{I}_n$) are monotonically decreasing. Omitting the iteration index for simplicity, from (4.68), (4.69), and

$$(\delta_{1,i} - S_{1,i})W_i = 1 + \sum_{j \neq i}\frac{W_j}{z_i - z_j}$$

(see (3.72)), we find

$$|\widehat{C}_i| < \frac{5}{3}|\widehat{W}_i| < \frac{5}{3} \cdot \frac{1}{4}|W_i| = \frac{5}{12}|L_i(\delta_{1,i} - S_{1,i})^{-1}|\left|\frac{(\delta_{1,i} - S_{1,i})W_i}{L_i}\right|$$

$$= \frac{5}{12}|C_i|\left|\frac{1}{L_i}\left(1 + \sum_{j \neq i}\frac{W_j}{z_i - z_j}\right)\right|,$$

i.e.,

$$|\widehat{C}_i| < \frac{5}{12}|x_i||C_i|, \quad \text{where} \quad x_i = \frac{1}{L_i}\left(1 + \sum_{j \neq i}\frac{W_j}{z_i - z_j}\right). \tag{4.71}$$

Using the inclusion (i) of Lemma 4.3 and the inequality

$$\left|1 + \sum_{j \neq i}\frac{W_j}{z_i - z_j}\right| \leq 1 + \sum_{j \neq i}\frac{|W_j|}{|z_i - z_j|} \leq 1 + \frac{(n-1)w}{d} < 1 + (n-1)c_n,$$

we find

$$x_i \in \left(1 + \sum_{j \neq i}\frac{W_j}{z_i - z_j}\right)\{1; q_n\},$$

so that, using (1.71),

$$|x_i| < (1 + (n-1)c_n)(1 + q_n) < \frac{22}{15}.$$

From (4.71), we now obtain

$$|\widehat{C}_i| < \frac{5}{12} \cdot \frac{22}{15}|C_i| = \frac{11}{18}|C_i|.$$

Therefore, the constant β which appears in Theorem 3.1 is equal to $\beta = 11/18 \approx 0.6111$. By induction, we prove that the inequalities

$$|C_i^{(m+1)}| < \frac{11}{18}|C_i^{(m)}|$$

hold for each $i = 1, \ldots, n$ and $m = 0, 1, \ldots$.

The quantity $g(\beta)$ appearing in (ii) of Theorem 3.1 is equal to $g(11/18) = 1/(1 - 11/18) = 18/7$. It remains to prove the disjunctivity of the inclusion disks

$$S_1 = \left\{z_1^{(0)}; \frac{18}{7}|C_1^{(0)}|\right\}, \ldots, S_n = \left\{z_n^{(0)}; \frac{18}{7}|C_n^{(0)}|\right\}$$

(assertion (ii) of Theorem 3.1). By virtue of (4.69), we have

$$|C_i^{(0)}| < \frac{5}{3}w^{(0)}$$

for all $i = 1, \ldots, n$. If we choose the index $k \in I_n$ such that

$$|C_k^{(0)}| = \max_{1 \leq i \leq n} |C_i^{(0)}|,$$

then

$$d^{(0)} > (3n + 2)w^{(0)} > \frac{3}{5}(3n + 2)|C_k^{(0)}| \geq \frac{3(3n + 2)}{10}\left(|C_i^{(0)}| + |C_j^{(0)}|\right)$$
$$> g(11/18)\left(|C_i^{(0)}| + |C_j^{(0)}|\right),$$

since

$$\frac{3(3n + 2)}{10} \geq 3.3 > g(11/18) = 2.571...$$

for all $n \geq 3$. This means that

$$|z_i^{(0)} - z_j^{(0)}| \geq d^{(0)} > g(11/18)(|C_i^{(0)}| + |C_j^{(0)}|) = \text{rad } S_i + \text{rad } S_j.$$

Hence, according to (1.69), it follows that the inclusion disks $S_1, \ldots, S_n$ are disjoint, which completes the proof of Theorem 4.6. $\square$

We conclude this section with the remark that the quantity c_n appearing in the initial condition (4.67) may be greater than $1/(3n + 2)$ for some particular methods belonging to the family (4.55). The use of a smaller c_n in Theorem 4.6 is the price that one usually has to pay for being more general.

4.4 Numerical Examples

To demonstrate the convergence speed and the behavior of some methods belonging to the families presented in Sects. 4.2 and 4.3, we have tested these methods in the examples of algebraic polynomials using multiprecision arithmetic. We have used several values for the parameter α and taken common starting approximations $z_1^{(0)}, \ldots, z_n^{(0)}$ for each method. The accuracy of approximations has been estimated by the maximal error

$$u^{(m)} = \max_{1 \leq i \leq n} |z_i^{(m)} - \zeta_i|,$$

where $m = 0, 1, \ldots$ is the iteration index. The stopping criterion has been given by the inequality

$$E^{(m)} = \max_{i \leq i \leq n} |P(z_i^{(m)})| < \tau,$$

where τ is a given tolerance.

In our numerical experiments, we have often used the fact that all zeros of a polynomial $P(z) = a_n z^n + a_{n-1} z^{n-1} + \cdots + a_1 z + a_0$ $(a_0, a_n \neq 0)$ lie inside the annulus

$$\{z \in \mathbb{C} : r < |z| < R\},$$
$$r = \frac{1}{2} \min_{1 \leq k \leq n} \left|\frac{a_0}{a_k}\right|^{1/k}, \quad R = 2 \max_{1 \leq k \leq n} \left|\frac{a_{n-k}}{a_n}\right|^{1/k} \qquad (4.72)$$

(see [57, Theorem 6.4b, Corollary 6.4k] and Sect. 1.2). All tested methods have started with Aberth's initial approximations [1]

$$z_k^{(0)} = -\frac{a_{n-1}}{n} + r_0 \exp(i\theta_k), \quad i = \sqrt{-1}, \quad \theta_k = \frac{\pi}{n}\left(2k - \frac{3}{2}\right) \quad (k = 1, \ldots, n). \qquad (4.73)$$

We content ourselves with three examples.

Example 4.1. Methods from the class (4.18) have been tested in the example of the monic polynomial P of degree $n = 25$ given by

$$\begin{aligned}
P(z) = {}& z^{25} + (0.752 + 0.729i)z^{24} + (-0.879 - 0.331i)z^{23} + (0.381 - 0.918i)z^{22} \\
& + (0.781 - 0.845i)z^{21} + (-0.046 - 0.917i)z^{20} + (0.673 + 0.886i)z^{19} \\
& + (0.678 + 0.769i)z^{18} + (-0.529 - 0.874i)z^{17} + (0.288 + 0.095i)z^{16} \\
& + (-0.018 + 0.799i)z^{15} + (-0.957 + 0.386i)z^{14} + (0.675 - 0.872i)z^{13} \\
& + (0.433 - 0.562i)z^{12} + (-0.760 + 0.128i)z^{11} + (-0.693 - 0.882i)z^{10} \\
& + (0.770 - 0.467i)z^9 + (-0.119 + 0.277i)z^8 + (0.274 - 0.569i)z^7 \\
& + (-0.028 - 0.238i)z^6 + (0.387 + 0.457i)z^5 + (-0.855 - 0.186i)z^4 \\
& + (0.223 - 0.048i)z^3 + (0.317 + 0.650i)z^2 + (-0.573 + 0.801i)z \\
& + (0.129 - 0.237i).
\end{aligned}$$

The coefficients $a_k \in \mathbb{C}$ of P (except the leading coefficient) have been chosen by the random generator as $Re(a_k) = \texttt{random(x)}$, $Im(a_k) = \texttt{random(x)}$, where $\texttt{random(x)} \in (-1, 1)$. Using (4.72), we find that all zeros of the above polynomial lie in the annulus $\{z : r = 0.3054 < |z| < 2.0947 = R\}$.

For comparison, we have also tested the well-known Durand–Kerner's method (D–K for brevity)

$$z_i^{(m+1)} = z_i^{(m)} - \frac{P(z_i^{(m)})}{\displaystyle\prod_{\substack{j=1 \\ j \neq i}}^{n} (z_i^{(m)} - z_j^{(m)})} \qquad (i \in I_n;\ m = 0, 1, \ldots) \qquad (4.74)$$

(see Comment (M_1) in Sect. 1.1). This method is one of the most efficient methods for the simultaneous approximation of all zeros of a polynomial. All tested methods started with Aberth's initial approximations given by (4.73) with $n = 25$, $a_{n-1} = 0.752 + 0.729\,i$. In this example, the stopping criterion was given by

$$E^{(m)} = \max_{1 \leq i \leq 25} |P(z_i^{(m)})| < \tau = 10^{-7}.$$

We have performed three experiments taking $r_0 = 1.2$, 10, and 100 in (4.73). The first value is equal to the arithmetic mean of the radii $r = 0.3054$ and $R = 2.0947$ of the inclusion annulus given above. The values $r_0 = 10$ and $r_0 = 100$ have been chosen to exhibit the influence of r_0 to the convergence speed of the tested methods but also to show very good convergence behavior in the situation when the initial approximations are very crude and considerably far from the sought zeros.

Table 4.1 gives the number of iterative steps for the considered iterative procedures (4.18) and the Durand–Kerner's method (4.74). From this table, we see that the fourth-order methods (4.18) require less than half of the iterations produced by the second-order methods (4.74) if the parameter α in (4.18) is not too large. This means that the convergence behavior of the proposed methods (4.18) is at least as good as the behavior of the Durand–Kerner's method (4.74).

Table 4.1 Family (4.18): the number of iterations for different $|z_i^{(0)}| = r_0$

r_0 \ α	0	1	−1	$1/(n-1)$	1,000	D–K (4.74)
1.2	8	8	5	11	7	13
10	24	28	24	22	36	65
100	40	56	49	39	62	124

Example 4.2. Methods from the class (4.55) have been tested in the example of the monic polynomial P of degree $n = 15$ given by

$$P(z) = z^{15} + (-0.732 + 0.921\,\mathrm{i})z^{14} + (0.801 - 0.573\,\mathrm{i})z^{13} + (0.506 - 0.713\,\mathrm{i})z^{12}$$
$$+ (-0.670 + 0.841\,\mathrm{i})z^{11} + (-0.369 - 0.682\,\mathrm{i})z^{10} + (0.177 - 0.946\,\mathrm{i})z^9$$
$$+ (-0.115 + 0.577\,\mathrm{i})z^8 + (0.174 - 0.956\,\mathrm{i})z^7 + (-0.018 - 0.438\,\mathrm{i})z^6$$
$$+ (0.738 + 0.645\,\mathrm{i})z^5 + (-0.655 - 0.618\,\mathrm{i})z^4 + (0.123 - 0.088\,\mathrm{i})z^3$$
$$+ (0.773 + 0.965\,\mathrm{i})z^2 + (-0.757 + 0.109\,\mathrm{i})z + 0.223 - 0.439\,\mathrm{i}.$$

The coefficients $a_k \in \mathbb{C}$ of P (except the leading coefficient) were chosen by the random generator as in Example 4.1. Using (4.72), we find that all zeros of the above polynomial lie in the annulus $\{z \in \mathbb{C} : r = 0.477 < |z| < 2.353 = R\}$.

For comparison, we have also tested the Durand–Kerner's method (4.74) and the Ehrlich–Aberth's method (4.57). The Ehrlich–Aberth's method (briefly E–A method) has been tested for the purpose of comparison related to the methods from the family (4.55) obtained for very large α. All tested methods started with Aberth's initial approximations given by (4.73). In this example, the stopping criterion was given by

$$E^{(m)} = \max_{1 \le i \le 15} |P(z_i^{(m)})| < \tau = 10^{-12}.$$

We have performed several experiments taking $r_0 = 0.2$, 0.5, 1, 2, 4, 6, 8, and 100 to investigate the behavior of the tested methods for initial approximations of various magnitudes. Table 4.2 gives the number of iterative steps for the methods of the family (4.55), the D–K method (4.74), and the E–A method (4.57).

From Table 4.2, we see that the fourth-order methods (4.55) (excepting the Euler-like method ($\alpha = 1$) for some r_0) require less than half of iterative steps produced by the D–K method (4.74) if the parameter α in (4.55) is not too large. A hundred performed experiments, involving polynomials of various

Table 4.2 Family (4.55): the number of iterations for different $|z_i^{(0)}| = r_0$

r_0 \ α	0	1	−1	$1/(n-1)$	1,000	E–A (4.57)	D–K (4.74)
0.2	a	18	14	12	9	16	100^b
0.5	a	10	8	11	9	9	100^b
1	17	9	9	9	7	7	22
2	10	10	8	9	9	9	16
4	11	14	11	12	14	14	26
6	12	18	13	12	15	16	32
8	15	25	15	14	19	19	36
100	26	c	28	24	38	38	73

[a] The Ostrowski-like method oscillates giving some approximations of insufficient accuracy
[b] The Durand–Kerner's method converges to the exact zeros but in more than 100 iterations
[c] Some approximations are found with very high precision, but the remaining ones cannot be improved by the Euler-like method

degrees, have led to the same conclusion. These numerical results, as well as Examples 4.1 and 4.2, point that the convergence behavior of the proposed methods (4.55) is at least as good as the behavior of the D–K method (4.74). In addition, numerical experiments show that the E–A method (4.57) behaves almost the same as the methods from the family (4.55) for large α, which coincides with theoretical results (see Sect. 4.3 and Remark 4.8).

Let us note that certain convergence features of some tested methods are explained by comments given below Table 4.2. For instance, the Ostrowski-like method ($\alpha = 0$) and the D–K method (4.74) are not so efficient for initial approximations of small magnitudes. The D–K method attains the required stopping criterion but after a great number of iterations. On the other hand, the Ostrowski-like method provides very high accuracy for most approximations with a few iterations but whilst others reach a certain (insufficient) accuracy and cannot be improved in the continuation of iterative procedure; practically, this method "oscillates." It is interesting to note that Euler-like method ($\alpha = 1$) shows similar behavior but for initial approximations of large magnitude.

Most problems appearing in practical realization of the methods from the family (4.55) can be overcome by the choice of initial approximations in such a way that they lie inside the annulus $\{z \in \mathbb{C} \ : \ r < |z| < R\}$, where r and R are determined by (4.72). Our numerical experiments with such a choice of initial approximations showed that all tested methods have almost the same convergence behavior for a wide range of values of the parameter α and very fast convergence.

Example 4.3. We tested the family of simultaneous methods (4.63) for multiple zeros in the example of the polynomial

$$
\begin{aligned}
P(z) &= (z - 1.9)^2(z - 2)^2(z - 2.1)^2(z^2 + 4z + 8)(z^2 + 1)^3 \\
&= z^{14} - 8z^{13} + 22.98z^{12} - 39.92z^{11} + 142.94z^{10} - 583.76z^9 + 1515.34z^8 \\
&\quad - 2867.92z^7 + 4412.62z^6 - 5380.4z^5 + 5251.53z^4 - 4340.48z^3 \\
&\quad + 2742.73z^2 - 1276.16z + 509.443.
\end{aligned}
$$

The exact zeros are $\zeta_1 = 1.9$, $\zeta_2 = 2$, $\zeta_3 = 2.1$, $\zeta_{4,5} = -2 \pm 2\,i$, and $\zeta_{6,7} = \pm i$ with respective multiplicities $\mu_1 = \mu_2 = \mu_3 = 2$, $\mu_4 = \mu_5 = 1$, and $\mu_{6,7} = 3$.

We terminated the iterative process when the stopping criterion

$$
E^{(m)} = \max_{1 \leq i \leq 7} |P(z_i^{(m)})| < \tau = 10^{-8}
$$

was satisfied. All tested methods started with Aberth's initial approximations (4.73) taking $n = 14$, $k = 7$, $a_{n-1} = -8$, and $r_0 = 1$; thus, we took initial approximations equidistantly spaced on the circle with radius $r_0 = 1$. The numbers of iterations are given below:

Ostrowski-like method,	$\alpha = 0$	12 iterations
Euler-like method,	$\alpha = 1$	7 iterations

Laguerre-like method, $\alpha = \mu_i/(n - \mu_i)$ 7 iterations
Halley-like method, $\alpha = -1$ 7 iterations
Large parameter method, $\alpha = 500$ 13 iterations

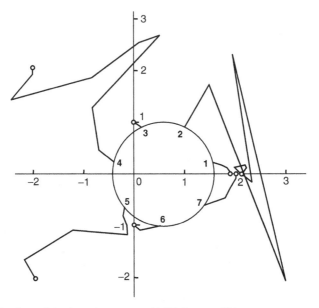

Fig. 4.1 The flow of the iterative process (4.63) for $\alpha = 500$

The flow of the iterative process (4.63) for $\alpha = 500$, concerning all zeros, is displayed in Fig. 4.1, where small circles represent the exact zeros. Let us note that the applied algorithm chooses itself the directions toward the sought zeros. This means that some approximations are good for the zeros in their vicinity (for instance, $z_3^{(0)}, z_6^{(0)}$), the others are rather far from the targets $(z_2^{(0)}, z_4^{(0)}, z_5^{(0)})$. In addition, there is a cluster of zeros $\{1.9, 2, 2.1\}$ in the considered example. However, despite these unsuitable conditions, the employed method overcomes aforementioned difficulties and finds all multiple zeros spending relatively little CPU time (i.e., it runs a small number of iterative steps).

Chapter 5
Families of Simultaneous Methods of Higher Order: Part II

In this chapter, we derive a fixed point relation of the square root type, which is the base for the construction of new one-parameter families of iterative methods for the simultaneous determination of simple complex zeros of a polynomial in ordinary complex arithmetic (Sect. 5.1) and circular complex arithmetic (Sect. 5.3). A slight modification of the derived fixed point relation can provide the simultaneous approximation of multiple zeros. Under computationally verifiable initial conditions, we prove that the basic method has the convergence order equal to 4. Using an approach with corrections, proposed by Carstensen and M. Petković in [17] and [111], we construct modified methods with very fast convergence on the account of only a few additional numerical operations (Sect. 5.2). In this way, we obtain a high computational efficiency of the proposed methods. Numerical results are given in Sect. 5.2 for the methods realized in ordinary complex arithmetic and in Sect. 5.3 for the methods implemented in circular complex arithmetic.

5.1 One-Parameter Family for Simple Zeros

In this section, we present a new family of high-order methods for the simultaneous determination of simple complex zeros of a polynomial. This family includes a complex parameter α ($\neq -1$) whose values will be discussed later. Let P be a monic polynomial with simple zeros $\zeta_1, \ldots, \zeta_n$ and let $z_1, \ldots, z_n$ be their mutually distinct approximations. For the point $z = z_i$ ($i \in I_n$), let us introduce the notations:

$$
\Sigma_{\lambda,i} = \sum_{\substack{j=1 \\ j \neq i}}^{n} \frac{1}{(z_i - \zeta_j)^\lambda}, \quad S_{\lambda,i} = \sum_{\substack{j=1 \\ j \neq i}}^{n} \frac{1}{(z_i - z_j)^\lambda} \quad (\lambda = 1, 2),
$$

$$
\delta_{1,i} = \frac{P'(z_i)}{P(z_i)}, \quad \Delta_i = \frac{P'(z_i)^2 - P(z_i)P''(z_i)}{P(z_i)^2}, \tag{5.1}
$$

M. Petković, *Point Estimation of Root Finding Methods*. Lecture Notes in Mathematics 1933,
© Springer-Verlag Berlin Heidelberg 2008

$$f_i^* = (\alpha + 1)\Sigma_{2,i} - \alpha(\alpha + 1)\Sigma_{1,i}^2, \quad f_i = (\alpha + 1)S_{2,i} - \alpha(\alpha + 1)S_{1,i}^2, \quad (5.2)$$

$$u_i = z_i - \zeta_i, \quad u = \max_{1 \le i \le n} |u_i|.$$

Lemma 5.1. *For $i \in I_n$, the following identity is valid*

$$(\alpha + 1)\Delta_i - \alpha\delta_{1,i}^2 - f_i^* = \left(\frac{\alpha + 1}{u_i} - \alpha\delta_{1,i}\right)^2. \tag{5.3}$$

Proof. Starting from the identities

$$\delta_{1,i} = \frac{P'(z_i)}{P(z_i)} = \sum_{j=1}^{n} \frac{1}{z_i - \zeta_j} = \frac{1}{u_i} + \Sigma_{1,i} \tag{5.4}$$

and

$$\Delta_i = \frac{P'(z_i)^2 - P(z_i)P''(z_i)}{P(z_i)^2} = -\left(\frac{P'(z_i)}{P(z_i)}\right)' = \sum_{j=1}^{n} \frac{1}{(z_i - \zeta_j)^2} = \frac{1}{u_i^2} + \Sigma_{2,i}, \tag{5.5}$$

we obtain

$$(\alpha + 1)\Delta_i - \alpha\delta_{1,i}^2 - f_i^* = (\alpha + 1)\left(\frac{1}{u_i^2} + \Sigma_{2,i}\right) - \alpha\left(\frac{1}{u_i} + \Sigma_{1,i}\right)^2$$

$$-(\alpha + 1)\Sigma_{2,i} + \alpha(\alpha + 1)\Sigma_{1,i}^2$$

$$= \frac{1}{u_i^2} - \frac{2\alpha}{u_i}\Sigma_{1,i} + \alpha^2\Sigma_{1,i}^2 - \frac{2\alpha}{u_i}\Sigma_{1,i}$$

$$= \frac{1}{u_i^2} - \frac{2\delta_{1,i}}{u_i}\left(\delta_{1,i} - \frac{1}{u_i}\right) + \alpha^2\left(\delta_{1,i} - \frac{1}{u_i}\right)^2$$

$$= \left(\frac{\alpha + 1}{u_i} - \alpha\delta_{1,i}\right)^2. \quad \square$$

The identity (5.3) is convenient for the construction of a fruitful fixed point relation. Solving for ζ_i yields

$$\zeta_i = z_i - \frac{\alpha + 1}{\alpha\delta_{1,i} + \left[(\alpha + 1)\Delta_i - \alpha\delta_{1,i}^2 - f_i^*\right]^{1/2}} \quad (i \in I_n), \tag{5.6}$$

assuming that two values of the square root have to be taken in (5.6).

To save the space, it is convenient to use a more compact notation as follows:

1° The approximations $z_1^{(m)}, \ldots, z_n^{(m)}$ of the zeros at the mth iterative step will be briefly denoted by $z_1, \ldots, z_n$, and the new approximations $z_1^{(m+1)}, \ldots, z_n^{(m+1)}$, obtained in the subsequent iteration by some simultaneous iterative method, by $\hat{z}_1, \ldots, \hat{z}_n$, respectively.

2° $S_{k,i}(\boldsymbol{a}, \boldsymbol{b}) = \sum_{j=1}^{i-1} \dfrac{1}{(z_i - a_j)^k} + \sum_{j=i+1}^{n} \dfrac{1}{(z_i - b_j)^k}$,

$f_i(\boldsymbol{a}, \boldsymbol{b}) = (\alpha + 1) S_{2,i}(\boldsymbol{a}, \boldsymbol{b}) - \alpha(\alpha + 1) S_{1,i}^2(\boldsymbol{a}, \boldsymbol{b})$,

where $\boldsymbol{a} = (a_1, \ldots, a_n)$ and $\boldsymbol{b} = (b_1, \ldots, b_n)$ are some vectors of distinct complex numbers. If $\boldsymbol{a} = \boldsymbol{b} = \boldsymbol{z} = (z_1, \ldots, z_n)$, then we will write $S_{k,i}(\boldsymbol{z}, \boldsymbol{z}) = S_{k,i}$ and $f_i(\boldsymbol{z}, \boldsymbol{z}) = f_i$ as in (5.2). Dealing with f_i^*, we always have $\boldsymbol{a} = \boldsymbol{b} = \boldsymbol{\zeta} = (\zeta_1, \ldots, \zeta_n)$ so that we will write only f_i^* for brevity, omitting arguments.

3° $\boldsymbol{z} = (z_1, \ldots, z_n)$ (the current vector of approximations),

$\hat{\boldsymbol{z}} = (\hat{z}_1, \ldots, \hat{z}_n)$ (the new vector of approximations).

Putting $\zeta_i := \hat{z}_i$ in (5.6), where $\hat{z}_i$ is a new approximation to the zero ζ_i, and taking certain approximations of ζ_j on the right side of the fixed point relation (5.6), we obtain a new one-parameter family of iterative methods for the simultaneous determination of all simple zeros of a polynomial. Omitting f_i^* in (5.6) and setting $\zeta_i := \hat{z}_i$, we observe that (5.6) reduces to the Hansen–Patrick's family (4.1). For this reason, all iterative methods derived in this chapter could be regarded as methods of Hansen–Patrick's type.

In our consideration of a new family, we will always assume that $\alpha \neq -1$. In the particular case $\alpha = -1$, the proposed family reduces (by applying a limit process) to the Halley-like method proposed by X. Wang and Zheng [184]. The corresponding initial conditions which ensure the guaranteed convergence of this method have been considered by M. Petković and Đ. Herceg [116].

For the total-step methods (parallel or the Jacobi mode) and the single-step methods (serial or the Gauss-Seidel mode), the abbreviations TS and SS will be used. First, we will construct the family of total-step methods [138]:

The basic total-step method (TS):

$$\hat{z}_i = z_i - \dfrac{\alpha + 1}{\alpha \delta_{1,i} + \left[(\alpha + 1)\Delta_i - \alpha \delta_{1,i}^2 - f_i(\boldsymbol{z}, \boldsymbol{z}) \right]_*^{1/2}} \qquad (i \in \boldsymbol{I}_n). \qquad \text{(TS)}$$

Remark 5.1. We assume that two values of the (complex) square root have to be taken in (5.6), (TS), and the modified iterative formulae presented later. As in Chap. 4, we will use the symbol $*$ to indicate the choice of the proper value of the square root, which appears in the presented iterative formulae. If approximations to the zeros are reasonably good, for instance, if the conditions (5.11) and (5.12) hold, then the sign $+$ should be taken.

Now, we present some special cases of the family (TS) taking various values of the parameter α:

$\alpha = 0$, *the Ostrowski-like method:*

$$\hat{z}_i = z_i - \frac{1}{\left[\Delta_i - S_{2,i}(z, z)\right]_*^{1/2}} \quad (i \in I_n). \tag{5.7}$$

$\alpha = 1/(n-1)$, *the Laguerre-like method:*

$$\hat{z}_i = z_i - \frac{n}{\delta_{1,i} + \left[(n-1)(n\Delta_i - \delta_{1,i}^2 - nS_{2,i}(z, z) + \dfrac{n}{n-1}S_{1,i}^2(z, z))\right]_*^{1/2}}$$
$$(i \in I_n). \tag{5.8}$$

$\alpha = 1$, *the Euler-like method:*

$$\hat{z}_i = z_i - \frac{2}{\delta_{1,i} + \left[2\Delta_i - \delta_{1,i}^2 - 2(S_{2,i}(z, z) - S_{1,i}^2(z, z))\right]_*^{1/2}} \quad (i \in I_n). \tag{5.9}$$

$\alpha = -1$, *the Halley-like method:*

$$\hat{z}_i = z_i - \frac{2\delta_{1,i}}{\Delta_i + \delta_{1,i}^2 - S_{2,i}(z, z) - S_{1,i}^2(z, z)} \quad (i \in I_n). \tag{5.10}$$

The Halley-like method (5.10) is obtained for $\alpha \to -1$ applying a limiting operation. This method was derived earlier by X. Wang and Zheng [182].

The names in (5.7)–(5.10) come from the similarity with the quoted classical methods. Indeed, omitting the sums $S_{1,i}$ and $S_{2,i}$ in the above formulae, we obtain the corresponding well-known classical methods (see Sect. 4.1).

Convergence Analysis

First, we give some estimates necessary for establishing the main convergence theorem. For simplicity, we will often omit the iteration index m and denote quantities in the subsequent $(m+1)$th iteration by $\hat{\ }$.

Using Corollary 1.1, we obtain the following upper error bounds.

Lemma 5.2. *Let $z_1 \ldots, z_n$ be distinct numbers satisfying the inequality $w < c_n d$, $c_n < 1/(2n)$. Then*

$$|u_i| = |\zeta_i - z_i| < \frac{|W_i|}{1 - nc_n} \quad (i \in I_n). \tag{5.11}$$

In this section, we assume that the following condition

$$w < c_n d, \qquad c_n = 1/(4n) \tag{5.12}$$

is fulfilled. Since $w < 1/(4n) < d/(2n)$, the assertions of Corollary 1.1 and Lemma 5.2 hold. According to Lemma 5.2, (5.12), and Corollary 1.1, we obtain that $D_1 = \{z_1; \frac{4}{3}|W_1|\}, \ldots, D_n = \{z_n; \frac{4}{3}|W_n|\}$ are mutually disjoint disks and each of them contains one and only one zero of the polynomial P. Also, in our convergence analysis, we will handle the parameter α lying in the disk $\{z : |z| < 1.8\}$ centered at the origin (i.e., $|\alpha| < 1.8$). These values for c_n and the upper bound of α have been found by using an extensive estimate-and-fitting procedure by using the programming package *Mathematica 6.0*.

Lemma 5.3. *Let $z_1, \ldots, z_n$ be distinct approximations to the zeros $\zeta_1, \ldots, \zeta_n$ and let $u_i = z_i - \zeta_i$ and $\hat{u}_i = \hat{z}_i - \zeta_i$, where $\hat{z}_1, \ldots, \hat{z}_n$ are approximations produced by the family of iterative methods (TS). If (5.12) holds and $|\alpha| < 1.8 \wedge \alpha \neq -1$, then*

$$|\hat{u}_i| \leq \frac{40n}{d^3}|u_i|^3 \sum_{j \neq i}|u_j| \quad (i \in I_n).$$

Proof. From (5.11), we obtain

$$|u_i| = |z_i - \zeta_i| < \frac{1}{1 - nc_n}|W_i| \leq \frac{1}{1 - nc_n}w < \frac{c_n}{1 - nc_n}d = \frac{d}{3n}. \qquad (5.13)$$

Then

$$|z_i - \zeta_j| \geq |z_i - z_j| - |z_j - \zeta_j| > d - \frac{c_n}{1 - nc_n}d = \frac{3n-1}{3n}d. \qquad (5.14)$$

Following the introduced notations, we find

$$f_i^* - f_i(z, z) = -(\alpha + 1) \sum_{j \neq i} \frac{u_j}{(z_i - z_j)(z_i - \zeta_j)}\left(\frac{1}{z_i - z_j} + \frac{1}{z_i - \zeta_j}\right)$$

$$+\alpha(\alpha + 1) \sum_{j \neq i} \frac{u_j}{(z_i - z_j)(z_i - \zeta_j)}\left(\sum_{j \neq i}\frac{1}{z_i - z_j} + \sum_{j \neq i}\frac{1}{z_i - \zeta_j}\right).$$

Using the bound (5.14) and the definition of the minimal distance d, we estimate

$$\frac{1}{|z_i - z_j||z_i - \zeta_j|} \leq \frac{3n}{(3n-1)d^2} =: a_n,$$

$$\frac{1}{|z_i - z_j|} + \frac{1}{|z_i - \zeta_j|} \leq \frac{1}{d} + \frac{3n}{(3n-1)d} = \frac{6n-1}{(3n-1)d} =: b_n. \qquad (5.15)$$

Then, from the previous expression for $f_i^* - f_i(z, z)$, we obtain for $i \in I_n$

$$|f_i^* - f_i(z, z)| \leq |\alpha + 1| b_n a_n \sum_{j \neq i} |u_j| + |\alpha||\alpha + 1| b_n a_n (n - 1) \sum_{j \neq i} |u_j|$$

$$\leq |\alpha + 1|(1 + |\alpha|(n - 1)) b_n a_n \sum_{j \neq i} |u_j|$$

$$< |\alpha + 1| \frac{(1 + 2|\alpha|)}{2} \frac{3n(n - 1)(6n - 1)}{(3n - 1)^2 d^3} \sum_{j \neq i} |u_j|. \qquad (5.16)$$

Since

$$\frac{\alpha + 1}{u_i} - \alpha \delta_{1,i} = \frac{\alpha + 1}{u_i} - \alpha \sum_{j=1}^n \frac{1}{z_i - \zeta_j} = \frac{1}{u_i} - \alpha \sum_{j \neq i} \frac{1}{z_i - \zeta_j},$$

by (5.13) and (5.14), we have

$$\left| \frac{\alpha + 1}{u_i} - \alpha \delta_{1,i} \right| \geq \frac{1}{|u_i|} - |\alpha| \sum_{j \neq i} \frac{1}{|z_i - \zeta_j|} > \frac{1}{|u_i|} - \frac{3n(n - 1)|\alpha|}{(3n - 1)d}$$

$$> \frac{1}{|u_i|} \left(1 - \frac{(n - 1)|\alpha|}{3n - 1} \right) > \frac{3 - |\alpha|}{3|u_i|} \qquad (5.17)$$

and

$$\left| \frac{\alpha + 1}{u_i} - \alpha \delta_{1,i} \right| |u_i| = \left| 1 - \alpha u_i \sum_{j \neq i} \frac{1}{z_i - \zeta_j} \right| \leq 1 + |\alpha||u_i| \sum_{j \neq i} \frac{1}{|z_i - \zeta_j|}$$

$$< 1 + \frac{3n(n - 1)}{(3n - 1)d} |\alpha||u_i| < \frac{3 + |\alpha|}{3}. \qquad (5.18)$$

Let us introduce the quantities y_i and v_i by

$$y_i = (\alpha + 1)\Delta_i - \alpha \delta_{1,i}^2 - f_i(z, z), \quad v_i = \frac{f_i^* - f_i(z, z)}{\left(\dfrac{\alpha + 1}{u_i} - \alpha \delta_{1,i} \right)^2}.$$

Using the identity (5.3), we obtain

$$y_i = (\alpha + 1)\Delta_i - \alpha \delta_{1,i}^2 - f_i^* + f_i^* - f_i(z, z) = \left(\frac{\alpha + 1}{u_i} - \alpha \delta_{1,i} \right)^2 (1 + v_i). \qquad (5.19)$$

According to the bounds (5.16) and (5.17) of $|f_i^* - f_i(z, z)|$ and $|(\alpha + 1)/ u_i - \alpha \delta_{1,i}|$, we estimate

$$|v_i| \leq \frac{|f_i^* - f_i(z, z)|}{|(\alpha + 1)/u_i - \alpha \delta_{1,i}|^2} < \frac{\dfrac{|\alpha + 1|(1 + 2|\alpha|)}{2} \dfrac{3n(n-1)(6n-1)}{(3n-1)^2 d^3} \sum_{j \neq i} |u_j|}{\left(\dfrac{3 - |\alpha|}{3|u_i|}\right)^2}$$

$$= \frac{|\alpha + 1|(1 + 2|\alpha|)}{(3 - |\alpha|)^2} \frac{27n(n-1)(6n-1)}{2(3n-1)^2} \frac{|u_i|^2}{d^3} \sum_{j \neq i} |u_j|$$

$$< \frac{|\alpha + 1|(1 + 2|\alpha|)}{(3 - |\alpha|)^2)} \frac{9n|u_i|^2}{d^3} \sum_{j \neq i} |u_j| =: h_i \quad (i \in I_n), \tag{5.20}$$

where we have used the inequality

$$\frac{6n - 1}{(3n - 1)^2} < \frac{2}{3(n - 1)} \quad \text{for all } n \geq 3.$$

Since $|u_i| < d/(3n)$ (according to (5.13)), we have

$$|u_i|^2 \sum_{j \neq i} |u_j| < \frac{(n - 1)d^3}{27n^3}, \tag{5.21}$$

so that (taking into account that $|\alpha| < 1.8$)

$$h_i < \frac{|\alpha + 1|(1 + 2|\alpha|)}{(3 - |\alpha|)^2} \frac{(n - 1)}{3n^2} \leq \frac{2(|\alpha| + 1)(1 + 2|\alpha|)}{27(3 - |\alpha|)^2} < 0.67 < 1 \quad (n \geq 3).$$

Therefore

$$\frac{1}{1 + \sqrt{1 - h_i}} < 0.64. \tag{5.22}$$

Since $|v_i| < h_i$, there follows $v_i \in V := \{0; h_i\}$, where V is the disk centered at the origin with the radius h_i. Now, using (1.64), (1.70), (1.72) (taking the principal branch), and the inequality (5.22), we find

$$\left[1 + v_i\right]_*^{1/2} \in \left[1 + \{0; h_i\}\right]_*^{1/2} = \left[\{1; h_i\}\right]_*^{1/2} = \left\{1; 1 - \sqrt{1 - h_i}\right\}$$

$$= \left\{1; \frac{h_i}{1 + \sqrt{1 - h_i}}\right\} \subset \left\{1; 0.64 h_i\right\}. \tag{5.23}$$

Let

$$x_i = \alpha \delta_{1,i} + \left[y_i\right]_*^{1/2}.$$

Then, using (5.18)–(5.20) and (5.23), as well as the properties (1.65), (1.70), and Theorem 1.7, we have

$$x_i = \alpha\delta_{1,i} + \left(\frac{\alpha+1}{u_i} - \alpha\delta_{1,i}\right)\left[1 + v_i\right]_*^{1/2}$$

$$\in \alpha\delta_{1,i} + \left(\frac{\alpha+1}{u_i} - \alpha\delta_{1,i}\right)\left\{1; 0.64h_i\right\}$$

$$= \left\{\frac{\alpha+1}{u_i}; \left|\frac{\alpha+1}{u_i} - \alpha\delta_{1,i}\right| |u_i|\frac{0.64h_i}{|u_i|}\right\}$$

$$\subset \left\{\frac{\alpha+1}{u_i}; \frac{0.64(3+|\alpha|)h_i}{3|u_i|}\right\}$$

$$= \left\{\frac{\alpha+1}{u_i}; |\alpha+1|\varphi(|\alpha|)\frac{1.92n|u_i|}{d^3}\sum_{j\neq i}|u_j|\right\},$$

where we put

$$\varphi(|\alpha|) = \frac{(1+2|\alpha|)(3+|\alpha|)}{(3-|\alpha|)^2}.$$

Since $\varphi(|\alpha|) < 15.5$ for $|\alpha| < 1.8$, we have

$$x_i \in \left\{\frac{\alpha+1}{u_i}; R_i\right\} \quad (i \in \mathbf{I}_n), \tag{5.24}$$

where

$$R_i = \frac{30n|\alpha+1||u_i|}{d^3}\sum_{j\neq i}|u_j| \quad (i \in \mathbf{I}_n). \tag{5.25}$$

Using (5.24), the centered inversion of a disk (1.63), (1.65), and (1.70), we obtain

$$\frac{\alpha+1}{x_i} \in \frac{\alpha+1}{\{(\alpha+1)/u_i; R_i\}} = (\alpha+1)\left\{\frac{u_i}{\alpha+1}; \frac{R_i}{\frac{|\alpha+1|}{|u_i|}\left(\frac{|\alpha+1|}{|u_i|} - R_i\right)}\right\}$$

$$= \left\{u_i; \frac{R_i|u_i|^2}{|\alpha+1| - R_i|u_i|}\right\}. \tag{5.26}$$

By (5.13), we find that $R_i|u_i| < \dfrac{20|\alpha+1|}{81}$ for all $n \geq 3$, so that

$$|\alpha+1| - R_i|u_i| > |\alpha+1| - \frac{20}{81}|\alpha+1| = \frac{61}{81}|\alpha+1| > 0.$$

Therefore, the disk in (5.26) is well defined. In addition, using (5.26) and the last inequalities, we get

$$\frac{\alpha+1}{x_i} \in \left\{ u_i; \frac{81R_i|u_i|^2}{61|\alpha+1|} \right\}. \tag{5.27}$$

The iterative formula (TS) can be written in the form

$$\hat{z}_i = z_i - \frac{\alpha+1}{x_i}, \tag{5.28}$$

wherefrom

$$\hat{z}_i - \zeta_i = z_i - \zeta_i - \frac{\alpha+1}{x_i}, \quad \text{i.e.,} \quad \hat{u}_i = u_i - \frac{\alpha+1}{x_i}.$$

Hence, by the inclusion (5.27), we find

$$\hat{u}_i = u_i - \frac{\alpha+1}{x_i} \in \left\{ 0; \frac{81R_i|u_i|^2}{61|\alpha+1|} \right\},$$

wherefrom, by (1.71) and (5.25),

$$|\hat{u}_i| \leq \frac{81}{61|\alpha+1|} R_i|u_i|^2 = \frac{81}{61|\alpha+1|} \frac{30n|\alpha+1||u_i|^3}{d^3} \sum_{j \neq i} |u_j|$$

$$< \frac{40n}{d^3} |u_i|^3 \sum_{j \neq i} |u_j| \quad (i \in I_n), \tag{5.29}$$

which proves Lemma 5.3. $\quad\square$

Lemma 5.4. *Under the condition of Lemma 5.3, the following assertions are valid for the iterative method* (TS):

(i) $d < \dfrac{9n}{9n-8}\hat{d}$.

(ii) $|\widehat{W}_i| < \frac{2}{3}|W_i|$.

(iii) $\hat{w} < \dfrac{d}{4n}$.

Proof. From (5.27) and (5.28), it follows

$$\hat{z}_i - z_i = -\frac{\alpha+1}{x_i} \in -\left\{ u_i; \frac{81R_i|u_i|^2}{61|\alpha+1|} \right\} \subset -\left\{ u_i; \frac{40n}{d^3}|u_i|^3 \sum_{j \neq i} |u_j| \right\}.$$

Using (1.71) and (5.13), from the last inclusion, we have

$$|\hat{z}_i - z_i| \leq |u_i| + \frac{40n}{d^3}|u_i|^3 \sum_{j \neq i} |u_j| \leq |u_i|\left(1 + \frac{40(n-1)}{27n^2}\right)$$

$$< \frac{4}{3}|u_i| < \frac{16}{9}w < \frac{4d}{9n}. \tag{5.30}$$

From (5.30), we obtain

$$|\hat{z}_i - z_j| \geq |z_i - z_j| - |\hat{z}_i - z_i| > d - \frac{4d}{9n} = \frac{9n-4}{9n}d \qquad (5.31)$$

and

$$|\hat{z}_i - \hat{z}_j| \geq |z_i - z_j| - |\hat{z}_i - z_i| - |\hat{z}_j - z_j| > d - \frac{8d}{9n} = \frac{9n-8}{9n}d. \qquad (5.32)$$

The inequality (5.32) yields

$$\hat{d} > \frac{9n-8}{9n}d, \quad \text{i.e.,} \quad \frac{d}{\hat{d}} < \frac{9n}{9n-8}, \qquad (5.33)$$

which proves (i) of the lemma.

By the inclusion (5.24), from the iterative formula (5.28), we obtain

$$\frac{1}{\hat{z}_i - z_i} = -\frac{x_i}{\alpha+1} \in \left\{ -\frac{1}{u_i}; \frac{R_i}{|\alpha+1|} \right\} = \left\{ -\frac{1}{u_i}; \frac{30n|u_i|}{d^3} \sum_{j \neq i} |u_j| \right\}$$

$$= \left\{ -\frac{1}{u_i}; \kappa_i \right\},$$

where we put $\kappa_i = \dfrac{30n|u_i|}{d^3} \sum_{j \neq i} |u_j|$. Hence

$$\frac{W_i}{\hat{z}_i - z_i} \in \left\{ -\frac{W_i}{u_i}; |W_i|\kappa_i \right\}. \qquad (5.34)$$

We use the identities (5.5) and (3.72) to find

$$\delta_{1,i} = \sum_{j=1}^{n} \frac{1}{z_i - \zeta_j} = \frac{1}{u_i} + \sum_{j \neq i} \frac{1}{z_i - \zeta_j} = \sum_{j \neq i} \frac{1}{z_i - z_j} + \frac{1}{W_i} \left(\sum_{j \neq i} \frac{W_j}{z_i - z_j} + 1 \right),$$

wherefrom

$$\frac{1}{u_i} = \sum_{j \neq i} \frac{1}{z_i - z_j} + \frac{1}{W_i} \left(\sum_{j \neq i} \frac{W_j}{z_i - z_j} + 1 \right) - \sum_{j \neq i} \frac{1}{z_i - \zeta_j}. \qquad (5.35)$$

Using (5.34) and (5.35), we get

$$\sum_{j=1}^{n} \frac{W_j}{\hat{z}_i - z_j} + 1 = \frac{W_i}{\hat{z}_i - z_i} + \sum_{j \neq i} \frac{W_j}{\hat{z}_i - z_j} + 1$$

$$\in \left\{ -\frac{W_i}{u_i}; |W_i|\kappa_i \right\} + \sum_{j \neq i} \frac{W_j}{\hat{z}_i - z_j} + 1$$

$$= \left\{ -\frac{W_i}{u_i} + \sum_{j \neq i} \frac{W_j}{\hat{z}_i - z_j} + 1; |W_i|\kappa_i \right\}$$

$$= \left\{ -W_i \sum_{j \neq i} \frac{1}{z_i - z_j} - \sum_{j \neq i} \frac{W_j}{z_i - z_j} - 1 + W_i \sum_{j \neq i} \frac{1}{z_i - \zeta_j} \right.$$

$$\left. + \sum_{j \neq i} \frac{W_j}{\hat{z}_i - z_j} + 1; |W_i|\kappa_i \right\}$$

$$= \{\Theta_i; \Psi_i\},$$

where

$$\Theta_i = -W_i \sum_{j \neq i} \frac{u_j}{(z_i - \zeta_j)(z_i - z_j)} - (\hat{z}_i - z_i) \sum_{j \neq i} \frac{W_j}{(\hat{z}_i - z_j)(z_i - z_j)} \qquad (5.36)$$

and

$$\Psi_i = |W_i|\kappa_i. \qquad (5.37)$$

Let us estimate the moduli of Θ_i and Ψ_i. Starting from (5.36) and using (5.12)–(5.14), (5.30), and (5.31), we find

$$|\Theta_i| \leq |W_i| \sum_{j \neq i} \frac{|u_j|}{|z_i - \zeta_j||z_i - z_j|} + |\hat{z}_i - z_i| \sum_{j \neq i} \frac{|W_j|}{|\hat{z}_i - z_j||z_i - z_j|}$$

$$< \frac{(n-1)\frac{4}{3}w^2}{\frac{3n-1}{3n}d^2} + \frac{16(n-1)w^2}{\frac{9(9n-4)}{9n}d^2} < \frac{n-1}{4n(3n-1)} + \frac{n-1}{n(9n-4)} < 0.05.$$

By virtue of (5.12) and (5.13), from (5.37), we obtain

$$\Psi_i = |W_i|\kappa_i < w\frac{30n|u_i|}{d^3} \sum_{j \neq i} |u_j| < w\frac{30n(n-1)}{d^3}\left(\frac{4w}{3}\right)^2$$

$$< \frac{5(n-1)}{6n^2} \leq \frac{5}{27} < 0.186. \qquad (5.38)$$

According to (1.71) and using the upper bounds of $|\Theta_i|$ and Ψ_i, we estimate

$$\left| \sum_{j=1}^{n} \frac{W_j}{\hat{z}_i - z_j} + 1 \right| < |\Theta_i| + \Psi_i < 0.05 + 0.186 < 0.236. \qquad (5.39)$$

Furthermore, using the bounds (5.30) and (5.32), we find

$$\left| \prod_{j \neq i} \frac{\hat{z}_i - z_j}{\hat{z}_i - \hat{z}_j} \right| \leq \prod_{j \neq i} \left(1 + \frac{|\hat{z}_j - z_j|}{|\hat{z}_i - \hat{z}_j|} \right) < \left(1 + \frac{4}{9n - 8} \right)^{n-1} < 1.56.$$

Having in mind the last inequality and the inequalities (5.30) and (5.32), we start from (3.71) for $z = \hat{z}_i$ and find

$$|\widehat{W}_i| = \left| \frac{P(\hat{z}_i)}{\prod_{j \neq i}(\hat{z}_i - \hat{z}_j)} \right| \leq |\hat{z}_i - z_i| \left| \sum_{j=1}^{n} \frac{W_j}{\hat{z}_i - z_j} + 1 \right| \left| \prod_{j \neq i} \frac{\hat{z}_i - z_j}{\hat{z}_i - \hat{z}_j} \right|$$

$$< \frac{16}{9}|W_i| \cdot 0.236 \cdot 1.56 < \frac{2}{3}|W_i|$$

and the assertion (ii) is proved.

According to (5.12), and (i) and (ii) of Lemma 5.3, we find for $|\alpha| < 1.8$

$$\hat{w} < \frac{2}{3}w < \frac{2}{3} \cdot \frac{d}{4n} < \frac{2}{12n} \cdot \frac{9n}{9n-8}\hat{d} < \frac{\hat{d}}{4n},$$

which completes the proof of the assertion (iii) of the lemma. $\quad\square$

Using results of Lemmas 5.3 and 5.4, we are able to state initial conditions which guarantee the convergence of the family of methods (TS) and to find its convergence order (see M. Petković and Rančić [138]).

Theorem 5.1. *Let P be a polynomial of degree $n \geq 3$ with simple zeros. If the initial condition*

$$w^{(0)} < c_n d^{(0)}, \quad c_n = \frac{1}{4n} \tag{5.40}$$

holds, then the family of simultaneous methods (TS) is convergent for $|\alpha| < 1.8$ with the order of convergence 4.

Proof. Using a similar technique presented in Chap. 3 and in the proofs of Lemmas 5.3 and 5.4, we prove the assertions of Theorem 5.1 by induction. Since (5.12) and (5.40) are of the same form, all estimates given in Lemmas 5.3 and 5.4 are valid for the index $m = 1$. Furthermore, the inequality (iii) in Lemma 5.4 coincides with (5.12), so that the assertions of Lemmas 5.3 and 5.4 are valid for the subsequent index, etc. Hence, by induction, we obtain the implication

$$w^{(m)} < c_n d^{(m)} \implies w^{(m+1)} < c_n d^{(m+1)}.$$

It involves the initial condition (5.40) under which all inequalities given in Lemmas 5.3 and 5.4 are valid for all $m = 0, 1, \ldots$. In particular, we have

$$\frac{d^{(m)}}{d^{(m+1)}} < \frac{9n}{9n-8} \tag{5.41}$$

and

$$|u_i^{(m+1)}| \leq \frac{40n}{\left(d^{(m)}\right)^3} |u_i^{(m)}|^3 \sum_{\substack{j=1 \\ j \neq i}}^{n} |u_j^{(m)}| \quad (i \in \boldsymbol{I}_n) \tag{5.42}$$

for each iteration index $m = 0, 1, \ldots$.

Let us substitute

$$t_i^{(m)} = \left[\frac{9n}{9n-8}(n-1)\frac{40n}{\left(d^{(m)}\right)^3} \right]^{1/3} |u_i^{(m)}|$$

in (5.42), then

$$t_i^{(m+1)} \leq \frac{9n-8}{9n(n-1)} \frac{d^{(m)}}{d^{(m+1)}} (t_i^{(m)})^3 \sum_{\substack{j=1 \\ j\neq i}}^{n} t_j^{(m)}.$$

Hence, by virtue of (5.41),

$$t_i^{(m+1)} \leq \frac{1}{n-1}(t_i^{(m)})^3 \sum_{\substack{j=1 \\ j\neq i}}^{n} t_j^{(m)} \quad (i \in \mathbf{I}_n, \; m = 0, 1, \ldots). \tag{5.43}$$

In regard to (5.13), for $|\alpha| < 1.8$, we find

$$t_i^{(0)} = \left[\frac{360n^2(n-1)}{(9n-8)(d^{(0)})^3} \right]^{1/3} |u_i^{(0)}| < \left[\frac{360n^2(n-1)}{(9n-8)(d^{(0)})^3} \right]^{1/3} \frac{d^{(0)}}{3n}$$

$$< \left[\frac{40(n-1)}{3n(9n-8)} \right]^{1/3} < 0.78 < 1.$$

Put $t = \max_i t_i^{(0)}$, then obviously $t_i^{(0)} \leq t < 1$ for all $i = 1, \ldots, n$ and $n \geq 3$. Hence, we conclude from (5.43) that the sequences $\{t_i^{(m)}\}$ (and, consequently, $\{|u_i^{(m)}|\}$) tend to 0 for all $i = 1, \ldots, n$. Therefore, $z_i^{(m)} \to \zeta_i$ ($i \in \mathbf{I}_n$) and the family of methods (TS) is convergent.

Using (5.30), (5.32), and (ii) of Lemma 5.4, we successively find

$$d^{(m)} > d^{(m-1)} - \frac{32}{9}w^{(m-1)} > d^{(m-2)} - \frac{32}{9}w^{(m-2)} - \frac{32}{9}w^{(m-1)}$$

$$\vdots$$

$$> d^{(0)} - \frac{32}{9}\left(w^{(0)} + w^{(1)} + \cdots + w^{(m-1)}\right)$$

$$> d^{(0)} - \frac{32}{9}w^{(0)}\left(1 + 2/3 + (2/3)^2 + \cdots + (2/3)^{m-1}\right)$$

$$> d^{(0)} - \frac{32}{3}w^{(0)} > d^{(0)} - \frac{32}{3}c_n d^{(0)} > \left(1 - \frac{8}{3n}\right)d^{(0)}.$$

Therefore, the quantity $1/d^{(m)}$ appearing in (5.42) is bounded by $1/d^{(m)} < 3n/((3n-8)d^{(0)})$, so that from (5.42) we conclude that the order of convergence of the total-step method (TS) is 4. $\square$

Remark 5.2. The condition $|\alpha| < 1.8$ is sufficient. This bound is used to ensure the validity of some (not so sharp) inequalities and estimates in the presented convergence analysis. However, in practice, we can take considerably larger value of $|\alpha|$, as many numerical examples have shown.

The convergence of the total-step method (TS) can be accelerated if we use new approximations to the zeros as soon as they are available (Gauss-Seidel approach). In this way, we obtain

The basic single-step method (SS):

$$\hat{z}_i = z_i - \frac{\alpha + 1}{\alpha\delta_{1,i} + \left[(\alpha+1)\Delta_i - \alpha\delta_{1,i}^2 - f_i(\hat{z}, z)\right]_*^{1/2}} \quad (i \in \boldsymbol{I}_n). \qquad \text{(SS)}$$

5.2 Family of Methods with Corrections

In this section, we state other modifications of the families (TS) and (SS) which possess very fast convergence. The proposed methods have a high computational efficiency since the acceleration of convergence is attained with only a few additional computations. Actually, the increase of the convergence rate is attained using Newton's and Halley's corrections, which use already calculated values of P, P', P'' at the points $z_1, \ldots, z_n$ – current approximations to the wanted zeros.

Let us introduce some notations:

$$N_i = N(z_i) = 1/\delta_{1,i} = \frac{P(z_i)}{P'(z_i)} \quad \text{(Newton's correction)}, \qquad (5.44)$$

$$H_i = H(z_i) = \left[\frac{P'(z_i)}{P(z_i)} - \frac{P''(z_i)}{2P'(z_i)}\right]^{-1} = \frac{2\delta_{1,i}}{\delta_{1,i}^2 + \Delta_i} \qquad (5.45)$$

$$\text{(Halley's correction)},$$

$$\boldsymbol{z}_N = (z_{N,1}, \ldots, z_{N,n}), \quad z_{N,i} = z_i - N(z_i) \quad \text{(Newton's approximations)},$$

$$\boldsymbol{z}_H = (z_{H,1}, \ldots, z_{H,n}), \quad z_{H,i} = z_i - H(z_i) \quad \text{(Halley's approximations)}.$$

We recall that the correction terms (5.44) and (5.45) appear in the iterative formulae

$$\hat{z} = z - N(z) \quad \text{(Newton's method)} \quad \text{and} \quad \hat{z} = z - H(z) \quad \text{(Halley's method)},$$

which have quadratic and cubic convergence, respectively.

Taking certain approximations z_j of ζ_j in the sums involved in f_i^* (see (5.2)) on the right side of the fixed point relation (5.6) and putting $\zeta_i := \hat{z}_i$ in (5.6), where $\hat{z}_i$ is a new approximation to the zero ζ_i, we obtain approximations f_i of f_i^*. Then from (5.6), we construct some improved families of iterative methods for the simultaneous determination of all simple zeros of a polynomial. First, we will construct the following families of total-step methods (parallel mode):

The basic total-step method (TS):

$$\hat{z}_i = z_i - \frac{\alpha + 1}{\alpha\delta_{1,i} + \left[(\alpha + 1)\Delta_i - \alpha\delta_{1,i}^2 - f_i(z, z)\right]_*^{1/2}} \quad (i \in I_n). \qquad \text{(TS)}$$

The total-step method with Newton's corrections (TSN):

$$\hat{z}_i = z_i - \frac{\alpha + 1}{\alpha\delta_{1,i} + \left[(\alpha + 1)\Delta_i - \alpha\delta_{1,i}^2 - f_i(z_N, z_N)\right]_*^{1/2}} \quad (i \in I_n). \quad \text{(TSN)}$$

The total-step method with Halley's corrections (TSH):

$$\hat{z}_i = z_i - \frac{\alpha + 1}{\alpha\delta_{1,i} + \left[(\alpha + 1)\Delta_i - \alpha\delta_{1,i}^2 - f_i(z_H, z_H)\right]_*^{1/2}} \quad (i \in I_n). \quad \text{(TSH)}$$

The corresponding single-step methods (serial mode) have the form:

The basic single-step method (SS):

$$\hat{z}_i = z_i - \frac{\alpha + 1}{\alpha\delta_{1,i} + \left[(\alpha + 1)\Delta_i - \alpha\delta_{1,i}^2 - f_i(\hat{z}, z)\right]_*^{1/2}} \quad (i \in I_n). \qquad \text{(SS)}$$

The single-step method with Newton's corrections (SSN):

$$\hat{z}_i = z_i - \frac{\alpha + 1}{\alpha\delta_{1,i} + \left[(\alpha + 1)\Delta_i - \alpha\delta_{1,i}^2 - f_i(\hat{z}, z_N)\right]_*^{1/2}} \quad (i \in I_n). \qquad \text{(SSN)}$$

The single-step method with Halley's corrections (SSH):

$$\hat{z}_i = z_i - \frac{\alpha + 1}{\alpha\delta_{1,i} + \left[(\alpha + 1)\Delta_i - \alpha\delta_{1,i}^2 - f_i(\hat{z}, z_H)\right]_*^{1/2}} \quad (i \in I_n). \qquad \text{(SSH)}$$

For some specific values of the parameter α, from the families of methods listed above, we obtain special cases such as the *Ostrowski-like method* ($\alpha = 0$, studied in [47] and [106]), the *Laguerre-like method* ($\alpha = 1/(n-1)$, considered in [134]), the *Euler-like method* ($\alpha = 1$), and the *Halley-like method* ($\alpha = -1$), see [177] and [182].

Convergence Analysis

Studying the convergence analysis of the total-step methods (TS), (TSN), and (TSH), we will investigate all three methods simultaneously. The same is valid for the single-step methods (SS), (SSN), and (SSH). For this purpose, we denote these methods with the additional superscript indices 1 (for (TS) and (SS)), 2 (for (TSN) and (SSN)), and 3 (for (TSH) and (SSH)) and, in the same manner, we denote the corresponding vectors of approximations as follows:

$$
\begin{aligned}
\boldsymbol{z}^{(1)} &= \boldsymbol{z} = (z_1, \dots, z_n), \\
\boldsymbol{z}^{(2)} &= \boldsymbol{z}_N = (z_{N,1}, \dots, z_{N,n}), \\
\boldsymbol{z}^{(3)} &= \boldsymbol{z}_H = (z_{H,1}, \dots, z_{H,n}).
\end{aligned}
$$

Now, we are able to present all the mentioned total-step methods (for $\alpha \neq -1$), denoted with (TS(k)) ($k = 1, 2, 3$), in the unique form as

$$
\hat{z}_i = z_i - \frac{\alpha + 1}{\alpha \delta_{1,i} + \left[(\alpha + 1)\Delta_i - \alpha \delta_{1,i}^2 - f_i(\boldsymbol{z}^{(k)}, \boldsymbol{z}^{(k)}) \right]_*^{1/2}} \qquad (i \in \boldsymbol{I}_n,\ k = 1, 2, 3).
$$

$$\text{(TS(k))}$$

Using the above notation for the arguments of f_i, the single-step methods (SS), (SSN), and (SSH), denoted commonly with (SS(k)), can be written in the unique form

$$
\hat{z}_i = z_i - \frac{\alpha + 1}{\alpha \delta_{1,i} + \left[(\alpha + 1)\Delta_i - \alpha \delta_{1,i}^2 - f_i(\hat{\boldsymbol{z}}, \boldsymbol{z}^{(k)}) \right]_*^{1/2}} \qquad (i \in \boldsymbol{I}_n,\ k = 1, 2, 3).
$$

$$\text{(SS(k))}$$

Computationally verifiable initial conditions which ensure the guaranteed convergence of the basic total-step method ($k = 1$) of Laguerre's type ($\alpha = 1/(n - 1)$) were established in [133].

Using the same technique presented in Sect. 5.1, we are able to state convergence theorems for the methods listed above. These theorems include the initial condition of the form $w < c_n d$ providing the guaranteed convergence, see [153] where this analysis is given in detail. However, this analysis occupies a lot of space so that we will give a convergence theorem under simplified conditions.

We assume that $\alpha \neq -1$ in all iterative formulae presented above. If $\alpha = -1$, then applying a limiting operation we obtain the methods of Halley's type

$$\hat{z}_i = z_i - \frac{2\delta_{1,i}}{\Delta_i + \delta_{1,i}^2 - S_{2,i}(z^{(k)}, z^{(k)}) - S_{1,i}^2(z^{(k)}, z^{(k)})} \quad (i \in I_n, \ k = 1, 2, 3),$$

whose basic variant and some improvements were considered in [109, Sect. 5.5], [177], and [182].

Let us introduce the notation

$$\eta = \min_{\substack{1 \le i, j \le n \\ i \ne j}} |\zeta_i - \zeta_j|, \quad q = \frac{4n}{\eta}$$

and suppose that the conditions

$$|u_i| = |z_i - \zeta_i| < \frac{\eta}{4n} = \frac{1}{q} \quad (i = 1, \ldots, n) \tag{5.46}$$

are satisfied. Also, in our convergence analysis, we will deal with the parameter α lying in the disk $|z| < 2.4$ centered at the origin (i.e., $|\alpha| < 2.4$).

Lemma 5.5. *Let* $z_1, \ldots, z_n$ *be distinct approximations to the zeros* $\zeta_1, \ldots, \zeta_n$ *and let* $u_i = z_i - \zeta_i$ *and* $\hat{u}_i = \hat{z}_i - \zeta_i$, *where* $\hat{z}_1, \ldots, \hat{z}_n$ *are approximations produced by the iterative methods* TS(k). *If (5.46) holds and* $|\alpha| < 2.4 \wedge \alpha \ne -1$, *then:*

(i) $|\hat{u}_i| \le \dfrac{q^{k+2}}{n-1} |u_i|^3 \sum_{j \ne i} |u_j|^k \quad (i \in I_n, \ k = 1, 2, 3).$

(ii) $|\hat{u}_i| < \dfrac{\eta}{4n} = \dfrac{1}{q} \quad (i \in I_n).$

The proof of the assertions (i) and (ii) is extensive but elementary, and can be derived applying a similar technique as the one used in [134]. The complete proof is given in [153]. For these reasons, we omit the proof.

Let $z_1^{(0)}, \ldots, z_n^{(0)}$ be approximations close enough to the zeros $\zeta_1, \ldots, \zeta_n$ of the polynomial P and let

$$u_i^{(m)} = z_i^{(m)} - \zeta_i, \quad u^{(m)} = \max_{1 \le i \le n} |u_i^{(m)}|,$$

where $z_1^{(m)}, \ldots, z_n^{(m)}$ are approximations obtained in the mth iterative step.

Theorem 5.2. *Let* $|\alpha| < 2.4 \wedge \alpha \ne -1$ *and let the inequalities*

$$|u_i^{(0)}| = |z_i^{(0)} - \zeta_i| < \frac{\eta}{4n} = \frac{1}{q} \quad (i = 1, \ldots, n) \tag{5.47}$$

hold. Then, the total-step methods (TS(k)) are convergent with the convergence order equal to $k + 3$ $(k = 1, 2, 3)$.

Proof. Starting from the condition (5.47) (which coincides with (5.46)) and using the assertion (i) of Lemma 5.5, we come to the following inequalities

$$|u_i^{(1)}| \leq \frac{q^{k+2}}{n-1}|u_i^{(0)}|^3 \sum_{\substack{j=1 \\ j \neq i}}^{n}|u_j^{(0)}|^k < \frac{1}{q} \quad (i \in \boldsymbol{I}_n, \ k = 1, 2, 3),$$

which means that the implications

$$|u_i^{(0)}| < \frac{\eta}{4n} = \frac{1}{q} \implies |u_i^{(1)}| < \frac{\eta}{4n} = \frac{1}{q} \quad (i \in \boldsymbol{I}_n)$$

are valid (see also the assertion (ii) of Lemma 5.5). We can prove by induction that the condition (5.47) implies

$$|u_i^{(m+1)}| \leq \frac{q^{k+2}}{n-1}|u_i^{(m)}|^3 \sum_{\substack{j=1 \\ j \neq i}}^{n}|u_j^{(m)}|^k < \frac{1}{q} \quad (i \in \boldsymbol{I}_n, \ k = 1, 2, 3) \qquad (5.48)$$

for each $m = 0, 1, \ldots$ and $i \in \boldsymbol{I}_n$. Replacing $|u_i^{(m)}| = t_i^{(m)}/q$ in (5.48), we get

$$t_i^{(m+1)} \leq \frac{\left(t_i^{(m)}\right)^3}{n-1} \sum_{\substack{j=1 \\ j \neq i}}^{n}\left(t_j^{(m)}\right)^k \quad (i \in \boldsymbol{I}_n, \ k = 1, 2, 3). \qquad (5.49)$$

Let $t^{(m)} = \max_{1 \leq i \leq n} t_i^{(m)}$. From (5.47), it follows

$$q|u_i^{(0)}| = t_i^{(0)} \leq t^{(0)} < 1 \quad (i \in \boldsymbol{I}_n).$$

Successive application of the inequalities of this type to (5.49) gives $t_i^{(m)} < 1$ for all $i \in \boldsymbol{I}_n$ and $m = 1, 2, \ldots$. According to this, we get from (5.49)

$$t_i^{(m+1)} \leq \left(t_i^{(m)}\right)^3\left(t^{(m)}\right)^k \leq \left(t^{(m)}\right)^{k+3} \quad (k = 1, 2, 3). \qquad (5.50)$$

From (5.50), we infer that the sequences $\{t_i^{(m)}\}$ $(i \in \boldsymbol{I}_n)$ converge to 0, which means that the sequences $\{|u_i^{(m)}|\}$ are also convergent, i.e., $z_i^{(m)} \to \zeta_i$ $(i \in \boldsymbol{I}_n)$. Finally, from (5.50), we may conclude that the total-step methods (TS(k)) have the convergence order $k + 3$, i.e., the total-step methods (TS) $(k = 1)$, (TSN) $(k = 2)$, and (TSH) $(k = 3)$ have the order of convergence *4*, *5*, and *6*, respectively. $\square$

Let us consider now the convergence rate of the single-step method (SS(k)). Applying the same technique and argumentations presented in [134] and starting from the initial conditions (5.47), we can prove that the inequalities

$$|u_i^{(m+1)}| \leq \frac{q^{k+2}}{n-1}|u_i^{(m)}|^3 \left(\sum_{j=1}^{i-1} |u_j^{(m+1)}| + q^{k-1} \sum_{j=i+1}^{n} |u_j^{(m)}|^k\right) < \frac{1}{q} \quad (k = 1, 2, 3)$$

(5.51)

hold for each $m = 0, 1, \ldots$ and $i \in I_n$, supposing that for $i = 1$ the first sum in (5.51) is omitted.

Substituting $|u_i^{(m)}| = t_i^{(m)}/q$ into (5.51), we obtain

$$t_i^{(m+1)} \leq \frac{\left(t_i^{(m)}\right)^3}{n-1} \left(\sum_{j=1}^{i-1} t_j^{(m+1)} + \sum_{j=i+1}^{n} \left(t_j^{(m)}\right)^k\right) \quad (i \in I_n, \ k = 1, 2, 3). \quad (5.52)$$

The convergence analysis of the single-step methods (SS(k)), similar to that presented by Alefeld and Herzberger [2], uses the notion of the R-order of convergence introduced by Ortega and Rheinboldt [98]. The R-order of an iterative process IP with the limit point ζ will be denoted by $O_R((\text{IP}), \zeta)$.

Theorem 5.3. *Assume that the initial conditions* (5.47) *and the inequalities* (5.51) *are valid for the single-step method* (SS(k)). *Then, the R-order of convergence of* (SS(k)) *is given by*

$$O_R((\text{SS(k)}), \zeta) \geq 3 + \tau_n(k), \quad (5.53)$$

where $\tau_n(k) > k$ *is the unique positive root of the equation*

$$\tau^n - k^{n-1}\tau - 3k^{n-1} = 0. \quad (5.54)$$

Proof. As in the proof of Theorem 5.2, we first note that the condition (5.47) implies

$$|u_i^{(0)}|q = t_i^{(0)} \leq t = \max_{1 \leq i \leq n} t_i^{(0)} < 1. \quad (5.55)$$

According to this and (5.52), we conclude that the sequences $\{t_i^{(m)}\}$ $(i \in I_n)$ converge to 0. Hence, the sequences $\{|u_i^{(m)}|\}$ are also convergent which means that $z_i^{(m)} \to \zeta_i$ $(i \in I_n)$.

Applying the technique by Alefeld and Herzberger [2], the following system of inequalities can be derived from the relations (5.52) and (5.55):

$$t_i^{(m+1)} \leq t^{s_i^{(m)}} \quad (i = 1, \ldots, n, \ m = 0, 1, \ldots). \quad (5.56)$$

The column vectors $s^{(m)} = [s_1^{(m)} \cdots s_n^{(m)}]^T$ are successively computed by

$$s^{(m+1)} = A_n(k) s^{(m)} \quad (5.57)$$

starting with $s^{(0)} = [1 \cdots 1]^T$. The $n \times n$-matrix $A_n(k)$ in (5.57) is given by

$$A_n(k) = \begin{bmatrix} 3\,k & & & & & \\ & 3\,k & & \mathbf{O} & & \\ & & 3 & k & & \\ & & & \ddots & \ddots & \\ & \mathbf{O} & & & 3 & k \\ 3\,k & 0 & 0 & \cdots & 0 & 3 \end{bmatrix} \qquad (k = 1, 2, 3)$$

(see [109, Sect. 2.3] for more general case). The characteristic polynomial of the matrix $A_n(k)$ is

$$g_n(\lambda; k) = (\lambda - 3)^n - (\lambda - 3)k^{n-1} - 3k^{n-1}.$$

Replacing $\tau = \lambda - 3$, we get

$$\phi_n(\tau; k) = g_n(\tau + 3; k) = \tau^n - k^{n-1}\tau - 3k^{n-1}.$$

It is easy to show that the equation

$$\tau^n - k^{n-1}\tau - 3k^{n-1} = 0$$

has the unique positive root $\tau_n(k) > k$. The corresponding (positive) eigenvalue of the matrix $A_n(k)$ is $3 + \tau_n(k)$. Using some elements of the matrix analysis, we find that the matrix $A_n(k)$ is irreducible and primitive, so that it has the unique positive eigenvalue equal to its spectral radius $\rho(A_n(k))$. According to the analysis presented by Alefeld and Herzberger [2], it can be shown that the spectral radius $\rho(A_n(k))$ gives the lower bound of the R-order of iterative method (SS(k)), for which the inequalities (5.53) are valid. Therefore, we have

$$O_R((SS(k)), \zeta) \geq \rho(A_n(k)) = 3 + \tau_n(k),$$

where $\tau_n(k) > k$ is the unique positive root of (5.54). $\square$

The lower bounds of $O_R((SS), \zeta)$, $O_R((SSN), \zeta)$, and $O_R((SSH), \zeta)$ are displayed in Table 5.1.

Table 5.1 The lower bound of the R-order of convergence

Methods \ n	3	4	5	6	7	10	20	Very large n
(SS)		4.67	4.45	4.34	4.27	4.23	4.15	4.07	$\rightarrow 4$
(SSN)		5.86	5.59	5.44	5.36	5.30	5.20	5.10	$\rightarrow 5$
(SSH)		6.97	6.66	6.50	6.40	6.34	6.23	6.11	$\rightarrow 6$

Methods for Multiple Zeros

Let us now consider a monic polynomial P of degree n with multiple zeros $\zeta_1, \ldots, \zeta_\nu$ $(\nu \leq n)$ of the respective multiplicities $\mu_1, \ldots, \mu_\nu$ $(\mu_1 + \cdots + \mu_\nu = n)$

$$P(z) = \prod_{j=1}^{\nu} (z - \zeta_j)^{\mu_j}.$$

For the point $z = z_i$ $(i \in I_\nu := \{1, \ldots, \nu\})$, let us rearrange the previous notations used for simple zeros:

$$\Sigma_{\lambda,i} = \sum_{\substack{j=1 \\ j \neq i}}^{\nu} \frac{\mu_j}{(z_i - \zeta_j)^\lambda}, \quad S_{\lambda,i} = \sum_{\substack{j=1 \\ j \neq i}}^{\nu} \frac{\mu_j}{(z_i - z_j)^\lambda} \quad (\lambda = 1, 2),$$

$$f_i^* = \mu_i(\alpha + 1)\Sigma_{2,i} - \alpha(\alpha + 1)\Sigma_{1,i}^2, \quad f_i = \mu_i(\alpha + 1)S_{2,i} - \alpha(\alpha + 1)S_{1,i}^2.$$

Lemma 5.6. *For $i \in I_\nu$, the following identity is valid*

$$\mu_i(\alpha + 1)\Delta_i - \alpha\delta_{1,i}^2 - f_i^* = \left(\frac{\mu_i(\alpha + 1)}{u_i} - \alpha\delta_{1,i}\right)^2. \tag{5.58}$$

Proof. Using the identities

$$\delta_{1,i} = \frac{P'(z_i)}{P(z_i)} = \sum_{j=1}^{\nu} \frac{\mu_j}{z_i - \zeta_j} = \frac{\mu_i}{u_i} + \sum_{\substack{j=1 \\ j \neq i}}^{\nu} \frac{\mu_i}{z_i - \zeta_j} \tag{5.59}$$

and

$$\Delta_i = \frac{P'(z_i)^2 - P(z_i)P''(z_i)}{P(z_i)^2} = -\left[\frac{d}{dz}\frac{P'(z)}{P(z)}\right]_{z=z_i} = \sum_{j=1}^{\nu} \frac{\mu_j}{(z_i - \zeta_j)^2}$$

$$= \frac{\mu_i}{u_i^2} + \sum_{\substack{j=1 \\ j \neq i}}^{\nu} \frac{\mu_j}{(z_i - \zeta_j)^2}, \tag{5.60}$$

we obtain

$$\mu_i(\alpha + 1)\Delta_i - \alpha\delta_{1,i}^2 - q_i^* = \mu_i(\alpha + 1)\left(\frac{\mu_i}{u_i^2} + \Sigma_{2,i}\right) - \alpha\left(\frac{\mu_i}{u_i} + \Sigma_{1,i}\right)^2$$

$$-\mu_i(\alpha + 1)\Sigma_{2,i} + \alpha(\alpha + 1)\Sigma_{1,i}^2$$

$$= \frac{\mu_i^2}{u_i^2} + \alpha^2\Sigma_{1,i}^2 - \frac{2\alpha\mu_i}{u_i}\Sigma_{1,i}$$

$$= \frac{\mu_i^2}{u_i^2} + \alpha^2 \left(\delta_{1,i}^2 + \frac{\mu_i^2}{u_i^2} - \frac{2\delta_{1,i}\mu_i}{u_i} \right) - \frac{2\alpha\mu_i}{u_i} \left(\delta_{1,i} - \frac{\mu_i}{u_i} \right)$$

$$= \left(\frac{\mu_i(\alpha+1)}{u_i} - \alpha\delta_{1,i} \right)^2. \quad \square$$

From the identity (5.58), we derive the following fixed point relation

$$\zeta_i = z_i - \frac{\mu_i(\alpha+1)}{\alpha\delta_{1,i} + \left[\mu_i(\alpha+1)\Delta_i - \alpha\delta_{1,i}^2 - f_i^* \right]^{1/2}} \quad (i \in I_\nu), \qquad (5.61)$$

assuming that two values of the square root have to be taken in (5.61). As in the previous chapters, the symbol $*$ indicates the choice of proper value. The fixed point relation (5.61) is suitable for the construction of iterative methods for the simultaneous finding of multiple zeros of a given polynomial [139]:

The total-step methods for multiple zeros:

$$\hat{z}_i = z_i - \frac{\mu_i(\alpha+1)}{\alpha\delta_{1,i} + \left[\mu_i(\alpha+1)\Delta_i - \alpha\delta_{1,i}^2 - f_i(z,z) \right]_*^{1/2}} \quad (i \in I_\nu). \qquad (5.62)$$

The single-step methods for multiple zeros:

$$\hat{z}_i = z_i - \frac{\mu_i(\alpha+1)}{\alpha\delta_{1,i} + \left[\mu_i(\alpha+1)\Delta_i - \alpha\delta_{1,i}^2 - f_i(\hat{z},z) \right]_*^{1/2}} \quad (i \in I_\nu). \qquad (5.63)$$

In the same manner as in Sect. 5.2, starting from (5.61) we can accelerate the iterative methods (5.62) and (5.63) using Schröder's and Halley's corrections given by

$$N(z_i) = \mu_i \frac{P(z_i)}{P'(z_i)} \quad \text{and} \quad H(z_i) = \frac{P(z_i)}{\left(\frac{1+1/\mu_i}{2} \right) P'(z_i) - \frac{P(z_i)P''(z_i)}{2P'(z_i)}}.$$

The order of convergence of the iterative methods (5.62) and (5.63) and their modifications with corrections is given through Theorems 5.2 and 5.3.

Numerical Results: Methods in Ordinary Complex Arithmetic

To compare the results of numerical experiments with theoretical predictions exposed in the preceding sections, a considerable number of polynomial equations were solved. To present approximations of very high accuracy, we implemented the corresponding algorithms using the programming package *Mathematica* 6.0 with multiprecision arithmetic.

As a measure of closeness of approximations to the exact zeros, we have calculated Euclid's norm

$$e^{(m)} := ||\mathbf{z}^{(m)} - \boldsymbol{\zeta}||_E = \left(\sum_{i=1}^{n} |z_i^{(m)} - \zeta_i|^2 \right)^{1/2}.$$

Example 5.1. Iterative methods (5.7)–(5.10) were applied for the simultaneous approximation to the zeros of the polynomial

$$\begin{aligned}
P(z) &= z^{11} - (4+\mathrm{i})z^{10} + (3+4\mathrm{i})z^9 - (38+3\mathrm{i})z^8 - (27-38\mathrm{i})z^7 \\
&\quad + (44+27\mathrm{i})z^6 + (117-44\mathrm{i})z^5 + (598-117\mathrm{i})z^4 - (934+598\mathrm{i})z^3 \\
&\quad + (360+934\mathrm{i})z^2 + (1800-360\mathrm{i})z - 1800\mathrm{i} \\
&= (z^2-4)(z^2+9)(z^2-2z+2)(z^2+2z+5)(z-\mathrm{i})(z+1)(z-5).
\end{aligned}$$

The exact zeros of this polynomial are ± 2, $\pm 3\mathrm{i}$, $1\pm\mathrm{i}$, $-1\pm 2\mathrm{i}$, i, -1, and 5. For the purpose of comparison, beside the methods (5.7)–(5.10), we also tested a particular method from the family (TS) which is obtained for $\alpha = 1/2$:

$$\hat{z}_i = z_i - \frac{3}{\delta_{1,i} + \sqrt{2}\left[3\Delta_i - \delta_{1,i}^2 - 3S_{2,i}(z,z) + \frac{3}{2}S_{1,i}^2(z,z) \right]_*^{1/2}}. \qquad (5.64)$$

Also, we applied the single-step version of (5.64).

All tested methods started with the following initial approximations:

$$z_1^{(0)} = 2.1 + 0.2\mathrm{i}, \quad z_2^{(0)} = -2.2 + 0.2\mathrm{i}, \quad z_3^{(0)} = 0.3 + 3.2\mathrm{i},$$

$$z_4^{(0)} = -0.2 - 3.2\mathrm{i}, \quad z_5^{(0)} = 1.2 + 1.2\mathrm{i}, \quad z_6^{(0)} = 0.7 - 0.8\mathrm{i},$$

$$z_7^{(0)} = -0.8 + 2.3\mathrm{i}, \quad z_8^{(0)} = -0.7 - 1.8\mathrm{i}, \quad z_9^{(0)} = -0.3 + 0.7\mathrm{i},$$

$$z_{10}^{(0)} = -1.2 + 0.2\mathrm{i}, \quad z_{11}^{(0)} = 4.8 + 0.3\mathrm{i}.$$

For these approximations, we have $e^{(0)} \approx 1.1$. The measure of accuracy $e^{(m)}$ ($m = 1, 2, 3$) is displayed in Table 5.2, where the denotation $A(-q)$ means $A \times 10^{-q}$.

Table 5.2 Euclid's norm of errors

Methods		$\alpha = 0$	$\alpha = \frac{1}{n-1}$	$\alpha = 1/2$	$\alpha = 1$	$\alpha = -1$
(TS)	$e^{(1)}$	$1.71(-2)$	$1.67(-2)$	$3.30(-2)$	$7.67(-2)$	$6.64(-2)$
	$e^{(2)}$	$4.17(-9)$	$3.74(-9)$	$8.95(-8)$	$2.51(-6)$	$2.38(-6)$
	$e^{(3)}$	$3.36(-35)$	$1.96(-35)$	$3.37(-30)$	$6.29(-24)$	$6.04(-24)$
(SS)	$e^{(1)}$	$2.31(-2)$	$2.07(-2)$	$1.98(-2)$	$4.18(-2)$	$5.79(-2)$
	$e^{(2)}$	$2.35(-9)$	$9.94(-10)$	$2.18(-9)$	$2.96(-7)$	$6.22(-7)$
	$e^{(3)}$	$2.16(-36)$	$1.80(-37)$	$7.20(-37)$	$6.96(-28)$	$1.37(-26)$

Example 5.2. Iterative methods (5.62) and (5.63), obtained for $\alpha = 0$, $\alpha = 1/(n-1)$, $\alpha = 1/2$, $\alpha = 1$, and $\alpha = -1$, were applied for the simultaneous determination of multiple zeros of the polynomial

$$P(z) = z^{13} - (1 - 2i)z^{12} - (10 + 2i)z^{11} - (30 + 18i)z^{10} + (35 - 62i)z^9$$
$$+ (293 + 52i)z^8 + (452 + 524i)z^7 - (340 - 956i)z^6$$
$$- (2505 + 156i)z^5 - (3495 + 4054i)z^4 - (538 + 7146i)z^3$$
$$+ (2898 - 5130i)z^2 + (2565 - 1350i)z + 675$$
$$= (z + 1)^4(z - 3)^3(z + i)^2(z^2 + 2z + 5)^2.$$

The exact zeros of this polynomial are $\zeta_1 = -1$, $\zeta_2 = 3$, $\zeta_3 = -i$, and $\zeta_{4,5} = -1 \pm 2i$ with respective multiplicities $\mu_1 = 4$, $\mu_2 = 3$, and $\mu_3 = \mu_4 = \mu_5 = 2$. The following complex numbers were chosen as starting approximations to these zeros:

$$z_1^{(0)} = -0.7 + 0.3i, \quad z_2^{(0)} = 2.7 + 0.3i, \quad z_3^{(0)} = 0.3 - 0.8i,$$
$$z_4^{(0)} = -1.2 - 2.3i, \quad z_5^{(0)} = -1.3 + 2.2i.$$

In the presented example for the initial approximations, we have $e^{(0)} \approx 1.43$. The measure of accuracy $e^{(m)}$ $(m = 1, 2, 3)$ is given in Table 5.3.

Table 5.3 Euclid's norm of errors

Methods		$\alpha = 0$	$\alpha = \frac{1}{n-1}$	$\alpha = 1/2$	$\alpha = 1$	$\alpha = -1$
(TS)	$e^{(1)}$	2.39(−2)	1.62(−2)	1.93(−2)	6.32(−2)	5.72(−2)
	$e^{(2)}$	1.47(−8)	1.18(−9)	1.39(−9)	8.80(−7)	1.54(−6)
	$e^{(3)}$	8.08(−34)	6.08(−38)	9.63(−38)	4.96(−26)	2.20(−26)
(SS)	$e^{(1)}$	1.54(−2)	1.38(−2)	1.42(−2)	1.51(−2)	1.99(−2)
	$e^{(2)}$	3.48(−10)	1.95(−10)	2.54(−10)	1.03(−9)	2.02(−9)
	$e^{(3)}$	1.18(−42)	2.35(−43)	1.19(−41)	5.72(−40)	2.40(−38)

Numerical experiments demonstrated very fast convergence of the modified methods with corrections. For illustration, we present the following numerical example.

Example 5.3. We applied the proposed methods (TS(k)) and (SS(k)) for the simultaneous determination of the zeros of the polynomial

$$P(z) = z^{11} + (1 - 4i)z^{10} - (6 + 4i)z^9 - (6 - 44i)z^8 - (36 - 44i)z^7$$
$$- (36 + 76i)z^6 + (186 - 76i)z^5 + (186 - 364i)z^4 - (445 - 364i)z^3$$
$$- (445 - 3600i)z^2 - (4500 - 3600i)z - 4500.$$

The exact zeros of this polynomial are $\zeta_1 = -1$, $\zeta_{2,3} = \pm 3$, $\zeta_4 = 5i$, $\zeta_{5,6} = \pm 2i$, $\zeta_{7,8} = 2 \pm i$, $\zeta_{9,10} = -2 \pm i$, and $\zeta_{11} = -i$.

The total-step as well as the single-step methods with Newton's and Halley's corrections, presented in Sect. 5.2, use the already calculated values P, P', P'' at the points $z_1, \ldots, z_n$, so that the convergence rate of these iterative methods is accelerated with negligible number of additional operations. Therefore, the employed approach provides the high computational efficiency of the proposed methods. Further decrease of the total number of operations may be attained by calculating the approximations $z_j - N(z_j)$ and $z_j - H(z_j)$ in advance, before summing the terms handling these approximations. In this way, the repeat calculations of the same quantities are avoided.

All tested methods started with the following initial approximations:

$$z_1^{(0)} = -1.2 - 0.3i, \ z_2^{(0)} = 3.3 + 0.2i, \quad z_3^{(0)} = -3.2 + 0.2i,$$
$$z_4^{(0)} = 0.3 + 4.8i, \quad z_5^{(0)} = 0.2 + 1.7i, \quad z_6^{(0)} = 0.2 - 2.2i,$$
$$z_7^{(0)} = 2.3 + 1.2i, \quad z_8^{(0)} = 1.8 - 0.7i, \quad z_9^{(0)} = -1.8 + 1.3i,$$
$$z_{10}^{(0)} = -1.8 - 0.8i, \ z_{11}^{(0)} = -0.2 - 0.8i.$$

The measure of accuracy $e^{(m)}$ ($m = 1, 2, 3$) is given in Tables 5.4 and 5.5 for the total-step and single-step methods, respectively. In the presented example for the initial approximations, we have $e^{(0)} = 1.11$.

Table 5.4 Euclid's norm of errors: total-step methods

Methods		(TS)	(TSN)	(TSH)
	$e^{(1)}$	2.88(−2)	1.72(−2)	5.53(−3)
$\alpha = 0$	$e^{(2)}$	6.71(−8)	9.91(−11)	1.25(−16)
	$e^{(3)}$	2.07(−30)	4.73(−53)	2.38(−99)
	$e^{(1)}$	2.68(−2)	1.70(−2)	5.47(−3)
$\alpha = \frac{1}{n-1}$	$e^{(2)}$	5.63(−8)	7.43(−11)	6.97(−17)
	$e^{(3)}$	3.70(−31)	1.39(−54)	1.25(−100)
	$e^{(1)}$	3.64(−2)	2.82(−2)	8.15(−3)
$\alpha = 1/2$	$e^{(2)}$	7.81(−8)	4.68(−10)	9.39(−15)
	$e^{(3)}$	1.84(−30)	4.55(−49)	4.70(−86)
	$e^{(1)}$	2.15(−1)	7.29(−2)	2.05(−2)
$\alpha = 1$	$e^{(2)}$	3.16(−4)	2.81(−7)	2.48(−11)
	$e^{(3)}$	1.30(−16)	5.42(−34)	5.06(−65)
	$e^{(1)}$	3.34(−1)	4.90(−2)	1.54(−2)
$\alpha = -1$	$e^{(2)}$	4.40(−4)	1.61(−8)	2.73(−13)
	$e^{(3)}$	7.17(−17)	1.68(−40)	3.62(−77)

From Tables 5.2–5.5 and a hundred tested polynomial equations, we can conclude that the results obtained by the proposed methods well match the theoretical results given in Theorems 5.1–5.3. Also, we note that two iterative steps of the presented families of methods are usually sufficient in solving most practical problems when initial approximations are reasonably good and polynomials are well conditioned. The third iteration is included to show remarkably fast convergence and give approximations of very high accuracy, rarely required in practice at present.

Table 5.5 Euclid's norm of errors: single-step methods

Methods		(SS)	(SSN)	(SSH)
$\alpha = 0$	$e^{(1)}$	$2.19(-2)$	$1.50(-2)$	$5.09(-3)$
	$e^{(2)}$	$6.60(-9)$	$1.62(-11)$	$9.90(-17)$
	$e^{(3)}$	$1.65(-37)$	$4.13(-60)$	$5.97(-104)$
$\alpha = \frac{1}{n-1}$	$e^{(1)}$	$2.18(-2)$	$1.52(-2)$	$4.98(-3)$
	$e^{(2)}$	$6.45(-9)$	$8.31(-12)$	$7.97(-17)$
	$e^{(3)}$	$3.36(-38)$	$6.70(-62)$	$1.14(-106)$
$\alpha = 1/2$	$e^{(1)}$	$3.46(-2)$	$2.52(-2)$	$7.41(-3)$
	$e^{(2)}$	$6.54(-8)$	$4.28(-10)$	$1.18(-15)$
	$e^{(3)}$	$1.89(-32)$	$1.60(-50)$	$1.44(-92)$
$\alpha = 1$	$e^{(1)}$	$2.14(-1)$	$5.32(-2)$	$1.81(-2)$
	$e^{(2)}$	$3.33(-4)$	$2.03(-8)$	$4.08(-12)$
	$e^{(3)}$	$2.61(-18)$	$1.89(-41)$	$1.71(-76)$
$\alpha = -1$	$e^{(1)}$	$5.04(-2)$	$3.12(-2)$	$9.88(-3)$
	$e^{(2)}$	$2.98(-7)$	$1.29(-9)$	$1.31(-14)$
	$e^{(3)}$	$2.89(-27)$	$6.60(-47)$	$7.12(-87)$

5.3 Family of Interval Methods

The fixed point relation (5.6) is suitable for the construction of interval iterative methods for the simultaneous inclusion of simple complex zeros of a polynomial. Let us assume that we have found mutually disjoint disks $Z_1, \ldots, Z_n$ with centers $z_i = \text{mid } Z_i$ and radii $r_i = \text{rad } Z_i$ such that $\zeta_i \in Z_i$ $(i \in \boldsymbol{I}_n)$. Let us substitute the zeros ζ_j by their inclusion disks Z_j in the expression for f_i^*, given by (5.2). In this way, we obtain a circular extension F_i of f_i^*

$$F_i = (\alpha + 1) \sum_{\substack{j=1 \\ j \neq i}}^{n} \left(\frac{1}{z_i - Z_j} \right)^2 - \alpha(\alpha + 1) \left(\sum_{\substack{j=1 \\ j \neq i}}^{n} \frac{1}{z_i - Z_j} \right)^2 \qquad (5.65)$$

with $f_i^* \in F_i$ for each $i \in \boldsymbol{I}_n$.

Using the inclusion isotonicity property (see Sect. 1.3), from the fixed point relation (5.6), we get

$$\zeta_i \in \hat{Z}_i := z_i - \frac{\alpha + 1}{\alpha \delta_{1,i} + \left[(\alpha + 1)\Delta_i - \alpha \delta_{1,i}^2 - F_i \right]^{1/2}} \qquad (i \in \boldsymbol{I}_n). \qquad (5.66)$$

If the denominator in (5.66) is a disk not containing 0, then $\hat{Z}_i$ is a new outer circular approximation to the zero ζ_i, i.e., $\zeta_i \in \hat{Z}_i$ $(i \in \boldsymbol{I}_n)$.

Let us introduce some notations:

1° The circular inclusion approximations $Z_1^{(m)}, \ldots, Z_n^{(m)}$ of the zeros at the mth iterative step will be briefly denoted by $Z_1, \ldots, Z_n$, and the new approximations $Z_1^{(m+1)}, \ldots, Z_n^{(m+1)}$, obtained by some simultaneous inclusion iterative method, by $\hat{Z}_1, \ldots, \hat{Z}_n$, respectively.

$2°$ $S_{k,i}(\boldsymbol{A}, \boldsymbol{B}) = \displaystyle\sum_{j=1}^{i-1}\left(\frac{1}{z_i - A_j}\right)^k + \sum_{j=i+1}^{n}\left(\frac{1}{z_i - B_j}\right)^k, \quad z_i = \operatorname{mid} Z_i,$

$\quad\quad F_i(\boldsymbol{A}, \boldsymbol{B}) = (\alpha + 1)S_{2,i}(\boldsymbol{A}, \boldsymbol{B}) - \alpha(\alpha + 1)S_{1,i}^2(\boldsymbol{A}, \boldsymbol{B}),$

where $\boldsymbol{A} = (A_1, \ldots, A_n)$ and $\boldsymbol{B} = (B_1, \ldots, B_n)$ are some vectors of disks. If $\boldsymbol{A} = \boldsymbol{B} = \boldsymbol{Z} = (Z_1, \ldots, Z_n)$, then we will sometimes write $S_{k,i}(\boldsymbol{Z}, \boldsymbol{Z}) = S_{k,i}$ and $F_i(\boldsymbol{Z}, \boldsymbol{Z}) = F_i$.

$3°$ $\boldsymbol{Z} = (Z_1, \ldots, Z_n)$ (the current disk approximations),

$\quad\quad \hat{\boldsymbol{Z}} = (\hat{Z}_1, \ldots, \hat{Z}_n)$ (the new disk approximations).

Starting from (5.66), we obtain a new one-parameter family of iterative methods for the simultaneous inclusion of all simple complex zeros of a polynomial. In our consideration of the new family, we will always suppose that $\alpha \neq -1$. However, the particular case $\alpha = -1$ reduces (by applying a limiting process) to the already known Halley-like interval method which was studied in [108], [182], [184], and [186].

First, following (5.66), we will construct the family of total-step methods [126]:

The basic interval total-step method (ITS):

$$\hat{Z}_i = z_i - \frac{\alpha + 1}{\alpha\delta_{1,i} + \left[(\alpha + 1)\Delta_i - \alpha\delta_{1,i}^2 - F_i(\boldsymbol{Z}, \boldsymbol{Z})\right]_*^{1/2}} \quad (i \in \boldsymbol{I}_n). \quad \text{(ITS)}$$

The symbol $*$ indicates that one of the two disks (say $U_{1,i} = \{c_{1,i}; d_i\}$ and $U_{2,i} = \{c_{2,i}; d_i\}$, where $c_{1,i} = -c_{2,i}$) has to be chosen according to a suitable criterion. That disk will be called a "proper" disk. From (5.3) and the inclusion $f_i^* \in F_i$, we conclude that the proper disk is the one which contains the complex number $(\alpha + 1)/u_i - \alpha\delta_{1,i}$. The choice of the proper sign in front of the square root in (ITS) was considered in detail in [47] (see also [109, Chap. 3]). The following criterion for the choice of the proper disk of a square root (between two disks) can be stated:

If the disks $Z_1, \ldots, Z_n$ are reasonably small, then we have to choose that disk (between $U_{1,i}$ and $U_{2,i}$) whose center minimizes

$$|P'(z_i)/P(z_i) - c_{k,i}| \quad (k = 1, 2).$$

The inclusion method (ITS) and its modifications which will be presented later are realized in circular complex interval arithmetic, which means that the produced approximations have the form of disks containing the wanted zeros. Therefore, these methods can be regarded as a self-validated numerical tool that features built-in upper error bounds to approximations expressed by the radii of the resulting disks. This enclosure property is the main advantage of inclusion methods.

Now, we present some special cases of the family (ITS) of iterative interval methods:

$\alpha = 0$, *the Ostrowski-like method:*

$$\widehat{Z}_i = z_i - \frac{1}{\left[\Delta_i - S_{2,i}(\boldsymbol{Z}, \boldsymbol{Z})\right]_*^{1/2}} \quad (i \in \boldsymbol{I}_n). \tag{5.67}$$

$\alpha = 1/(n-1)$, *the Laguerre-like method:*

$$\widehat{Z}_i = z_i - \frac{n}{\delta_{1,i} + \left[(n-1)\left(n\Delta_i - \delta_{1,i}^2 - nS_{2,i}(\boldsymbol{Z}, \boldsymbol{Z}) + \dfrac{n}{n-1}S_{1,i}^2(\boldsymbol{Z}, \boldsymbol{Z})\right)\right]_*^{1/2}}$$

$$(i \in \boldsymbol{I}_n). \tag{5.68}$$

$\alpha = 1$, *the Euler-like method:*

$$\widehat{Z}_i = z_i - \frac{2}{\delta_{1,i} + \left[2\Delta_i - \delta_{1,i}^2 - 2\left(S_{2,i}(\boldsymbol{Z}, \boldsymbol{Z}) - S_{1,i}^2(\boldsymbol{Z}, \boldsymbol{Z})\right)\right]_*^{1/2}} \quad (i \in \boldsymbol{I}_n).$$

$$\tag{5.69}$$

$\alpha = -1$, *the Halley-like method:*

$$\widehat{Z}_i = z_i - \frac{2\delta_{1,i}}{\Delta_i + \delta_{1,i}^2 - S_{2,i}(\boldsymbol{Z}, \boldsymbol{Z}) - S_{1,i}^2(\boldsymbol{Z}, \boldsymbol{Z})} \quad (i \in \boldsymbol{I}_n). \tag{5.70}$$

The Halley-like method is obtained for $\alpha \to -1$ applying a limiting operation.

The names come from the similarity with the quoted classical methods. For instance, omitting the sum in (5.67), we obtain the well-known Ostrowski's method $\hat{z}_i = z_i - 1/[\Delta_i]_*^{1/2}$, see [99].

For $m = 0, 1, 2, \dots$ and $n \geq 3$, let us introduce

$$r^{(m)} = \max_{1 \leq i \leq n} r_i^{(m)}, \quad \rho^{(m)} = \min_{\substack{1 \leq i,j \leq n \\ i \neq j}} \{|z_i^{(m)} - z_j^{(m)}| - r_j^{(m)}\}.$$

The quantity $\rho^{(m)}$ can be regarded as a measure of the separation of disks $Z_j^{(m)}$ from each other. The following assertion was proved by M. Petković and Milošević [126].

Theorem 5.4. *Let the interval sequences $\{Z_i^{(m)}\}$ $(i \in \boldsymbol{I}_n)$ be defined by the iterative formula* (ITS), *where $|a| < 1.13$. Then, under the condition*

$$\rho^{(0)} > 4(n-1)r^{(0)},$$

for each $i \in \boldsymbol{I}_n$ and $m = 0, 1, \dots$, we have:

(i) $\zeta_i \in Z_i^{(m)}$.

(ii) $r^{(m+1)} < \dfrac{14(n-1)^2\left(r^{(m)}\right)^4}{\left(\rho^{(0)} - \frac{17}{11}r^{(0)}\right)^3}$.

The above theorem asserts that (1) each of the produced disks contains the wanted zero in every iteration and (2) the convergence order of the interval method (ITS) is 4.

The total-step interval method (ITS) can be accelerated by using already calculated disk approximations in the current iteration (Gauss-Seidel approach). In this way, we obtain the single-step interval method:

$$\hat{Z}_i = z_i - \frac{\alpha+1}{\alpha\delta_{1,i} + \left[(\alpha+1)\Delta_i - \alpha\delta_{1,i}^2 - F_i(\hat{Z}, Z)\right]_*^{1/2}} \qquad (i \in I_n). \qquad \text{(ISS)}$$

The R-order of convergence of the single-step method (ISS) is given by Theorem 5.3 for $k = 1$ (the method without corrections).

The interval method (ITS) can be easily modified for the inclusion of multiple zeros (with known multiplicities) in a similar way as in Sect. 5.2. Starting from the fixed point relation (5.61), we obtain a new one-parameter family of iterative methods for the simultaneous inclusion of all multiple zeros of a polynomial:

$$\hat{Z}_i = z_i - \frac{\mu_i(\alpha+1)}{\alpha\delta_{1,i} + \left[\mu_i(\alpha+1)\Delta_i - \alpha\delta_{1,i}^2 - F_i\right]_*^{1/2}} \qquad (i \in I_\nu = \{1, \ldots, \nu\}),$$

$$\tag{5.71}$$

where

$$F_i = \mu_i(\alpha+1)S_{2,i} - \alpha(\alpha+1)S_{1,i}^2$$

$$= \mu_i(\alpha+1)\sum_{\substack{j=1 \\ j \neq i}}^{\nu} \mu_j\left(\frac{1}{z_i - Z_j}\right)^2 - \alpha(\alpha+1)\left(\sum_{\substack{j=1 \\ j \neq i}}^{\nu} \frac{\mu_j}{z_i - Z_j}\right)^2.$$

More details about this method, including the convergence analysis and numerical examples, can be found in [125] where the following convergence theorem was proved.

Theorem 5.5. *Let the interval sequences* $\left\{Z_i^{(m)}\right\}$ ($i \in I_\nu$) *be defined by the iterative formula* (5.71), *where* $|a| < 1.1$. *Then, under the condition*

$$\rho^{(0)} > 4(n - \mu)r^{(0)},$$

the following assertions hold for each $i \in \boldsymbol{I}_\nu$ and $m = 0, 1, \ldots$:

(i) $\zeta_i \in Z_i^{(m)}$.

(ii) $r^{(m+1)} < \dfrac{17(n - \mu)^2 \left(r^{(m)}\right)^4}{\mu \left(\rho^{(0)} - \frac{71}{43} r^{(0)}\right)^3}$.

Other modifications of the interval method (ITS) of the higher order that use Newton's and Halley's corrections (see Sect. 5.2) were studied by M. Petković and Milošević [127].

Numerical Results: Methods in Circular Complex Arithmetic

The presented family of inclusion methods and its modifications have been tested in solving many polynomial equations. In the implementation of these fast algorithms, we have applied a multistage *globally convergent* composite algorithm:

(a) *Find an inclusion region of the complex plane which includes all zeros of a polynomial.* It is well known that all zeros of a polynomial $P(z) = z^n + a_{n-1}z^{n-1} + \cdots + a_1 z + a_0$ lie in the disk centered at the origin with the radius

$$R = 2 \max_{1 \leq k \leq n} |a_{n-k}|^{1/k} \qquad (5.72)$$

(see Henrici [57, p. 457]). It is possible to use other similar formulae (see, e.g., [155]), but (5.72) has been found to be sufficient and satisfactory in practice.

(b) *Apply a slow convergent search algorithm to obtain mutually disjoint rectangles with a prescribed tolerance for the semidiagonals (sufficient to provide the convergence), each containing only one zero.* Let $C' = \bigcirc(S)$ and $S' = \square(C)$ denote a circle C' which circumscribes a square S and a square S' which circumscribes a circle C, respectively. Starting with the initial inclusion square $\square(\{0; R\})$, we use the efficient squaring subdividing procedure presented in [58] to find reasonably well-separated squares, say $S_1, \ldots, S_\nu$ ($\nu \leq n$). In this process, we employ appropriate inclusion tests (see [7], [57], [90], [100], [128], [155]) to determine which of the intermediate smaller squares still contains zeros, discarding all that do not. Thereafter we find initial inclusion disjoint disks $C_1 = Z_1^{(0)} = \bigcirc(S_1), \ldots, C_\nu = Z_\nu^{(0)} = \bigcirc(S_\nu)$.

(c) *Determine the number of zeros contained in each initial disk.* The methods based on the *principle of argument* and realized by numerical integration in the complex plane give satisfactory results in practice, see [46], [57], [65], [78], [79], [87], [93]. If the number of zeros, say $N(C_i)$, is greater than 1, then we take the center of each initial disk $z_i = \operatorname{mid} C_i$ to be a zero approximation and estimate the multiplicity of a (possibly multiple) zero contained in the disk C_i using the Lagouanelle formula (see [75], [77])

$$\mu_i = \lim_{z_i \to \zeta_i} \frac{P'(z_i)^2}{P'(z_i)^2 - P(z_i)P''(z_i)}.$$

For practical purpose, this formula is adapted for approximate calculation as follows

$$\mu_i = \mathcal{R}\left[\frac{P'(z_i)^2}{P'(z_i)^2 - P(z_i)P''(z_i)}\right], \tag{5.73}$$

where the denotation $\mathcal{R}[A_i]$ means that the calculated value A_i (in bracket) should be *rounded* to the nearest integer. In this estimating process, we assume that the radius of C_i is small enough to provide reasonably good approximation z_i to the zero ζ_i. To control the result, the order of multiplicity may also be determined by a method presented recently by Niu and Sakurai [93]. If (1) $\mu_i = N(C_i)$, we conclude that the disk C_i contains one zero of the multiplicity μ_i. If (2) $\mu_i \neq N(C_i)$ or $|A_i - \mathcal{R}[A_i]| > \tau$, where τ (> 0.1, say) is a tolerance which estimates the "goodness" of rounding in (5.73), then there is an indication that the disk C_i contains a cluster of $N(C_i)$ close zeros instead of one multiple zero. In this chapter, we consider only the case (1). Since the study of clusters is very actual, the reader is referred to the papers [7], [61], [78]–[80], [91], [94], [100], [155], [191] for more details.

(d) *Improve disks $C_1, \ldots, C_\nu$ containing the zeros of the multiplicities $\mu_1, \ldots, \mu_\nu$, respectively (determined in (c)), with rapidly convergent iterative methods to any required accuracy.*

Here, we present numerical results obtained by the interval method (ITS) described in Sect. 5.3. For comparison, beside the methods (5.67)–(5.70), we tested the method (ITS) which is obtained for $\alpha = 1/2$

$$\widehat{Z}_i = z_i - \frac{3}{\delta_{1,i} + \sqrt{2}\left[3\Delta_i - \delta_{1,i}^2 - 3S_{2,i}(\boldsymbol{Z}, \boldsymbol{Z}) + \frac{3}{2}S_{1,i}^2(\boldsymbol{Z}, \boldsymbol{Z})\right]_*^{1/2}}, \tag{5.74}$$

and the following inclusion methods of the fourth order

$$\widehat{Z}_i = z_i - \frac{1}{\dfrac{P'(z_i)}{P(z_i)} - \displaystyle\sum_{\substack{j=1 \\ j \neq i}}^{n} (z_i - Z_j + N_j)^{I_c}} \quad (i \in \boldsymbol{I}_n) \tag{5.75}$$

(see Carstensen and M. Petković [17]) and

$$\widehat{Z}_i = z_i - \frac{W(z_i)}{1 + \displaystyle\sum_{\substack{j=1 \\ j \neq i}}^{n} W(z_j)(Z_i - W_i - z_j)^{I_c}} \quad (i \in \boldsymbol{I}_n) \tag{5.76}$$

(see M. Petković and Carstensen [111]). We recall that the superscript index I_c denotes the centered inversion of a disk, see (1.63).

Some authors consider that it is better to apply more iterations of Weierstrass' method of the second order (see [109, Chap. 3])

$$\hat{Z}_i = z_i - P(z_i)\left(\prod_{\substack{j=1 \\ j \neq i}}^{n}(z_i - Z_j)\right)^{-1} \quad (i \in \boldsymbol{I}_n) \tag{5.77}$$

than any higher-order method. To check this opinion, we also tested this method.

Numerical experiments showed very fast convergence of the inclusion methods even in the case of relatively large initial disks. In all tested examples, the choice of initial disks was carried out under weaker conditions than those given in Theorem 5.4; moreover, the ratio $\rho^{(0)}/r^{(0)}$ was most frequently two, three, or more times less than $4(n-1)$. We have selected two typical examples.

Example 5.4. The interval methods (5.67)–(5.70) and (5.74)–(5.77) were applied for the simultaneous inclusion of the zeros of the polynomial

$$P(z) = z^9 + 3z^8 - 3z^7 - 9z^6 + 3z^5 + 9z^4 + 99z^3 + 297z^2 - 100z - 300.$$

The exact zeros of this polynomial are -3, ± 1, $\pm 2i$, and $\pm 2 \pm i$. The initial disks were selected to be $Z_i^{(0)} = \{z_i^{(0)}; 0.3\}$ with the centers

$$z_1^{(0)} = -3.1 + 0.2i, \ z_2^{(0)} = -1.2 - 0.1i, \ z_3^{(0)} = 1.2 + 0.1i,$$
$$z_4^{(0)} = 0.2 - 2.1i, \ \ z_5^{(0)} = 0.2 + 1.9i, \ \ z_6^{(0)} = -1.8 + 1.1i,$$
$$z_7^{(0)} = -1.8 - 0.9i, \ z_8^{(0)} = 2.1 + 1.1i, \ z_9^{(0)} = 1.8 - 0.9i.$$

The entries of the maximal radii of the disks produced in the first three iterations, for different values of α, are given in Table 5.6. We observe that the Weierstrass-like method (5.77) diverged.

Table 5.6 The maximal radii of inclusion disks

Methods	$r^{(1)}$	$r^{(2)}$	$r^{(3)}$
(ITS) $\alpha = 1$	1.96(−2)	5.32(−9)	7.95(−39)
(ITS) $\alpha = 1/2$	1.45(−2)	7.13(−10)	4.64(−43)
(ITS) $\alpha = \frac{1}{n-1}$	9.03(−3)	3.96(−10)	4.81(−42)
(ITS) $\alpha = 0$	8.09(−3)	3.20(−10)	1.70(−40)
(ITS) $\alpha = -1$	2.38(−2)	4.28(−8)	4.62(−34)
(5.75)	5.38(−2)	1.11(−5)	4.90(−23)
(5.76)	1.12(−2)	9.97(−9)	3.38(−34)
(5.77)	Diverges	–	–

Example 5.5. The same interval methods from Example 5.4 were applied for the determination of the eigenvalues of Hessenberg's matrix H (see Stoer

and Bulirsch [170]). Gerschgorin's disks were taken as the initial regions containing these eigenvalues. It is known that these disks are of the form $\{h_{ii}; R_i\}$ $(i = 1, \ldots, n)$, where h_{ii} are the diagonal elements of a matrix $[h_{ij}]_{n \times n}$ and $R_i = \sum_{j \neq i} |h_{ij}|$. If these disks are mutually disjoint, then each of them contains one and only one eigenvalue, which is very convenient for the application of inclusion methods.

The methods were tested in the example of the matrix

$$H = \begin{bmatrix} 2+3i & 1 & 0 & 0 & 0 \\ 0 & 4+6i & 1 & 0 & 0 \\ 0 & 0 & 6+9i & 1 & 0 \\ 0 & 0 & 0 & 8+12i & 1 \\ 1 & 0 & 0 & 0 & 10+15i \end{bmatrix},$$

whose characteristic polynomial is

$$g(\lambda) = \lambda^5 - (30 + 45i)\lambda^4 + (-425 + 1020i)\lambda^3 + (10350 - 2025i)\lambda^2$$
$$-(32606 + 32880i)\lambda - 14641 + 71640i.$$

We selected Gerschgorin's disks

$$Z_1 = \{2 + 3i; 1\}, \quad Z_2 = \{4 + 6i; 1\}, \quad Z_3 = \{6 + 9i; 1\},$$

$$Z_4 = \{8 + 12i; 1\}, \quad Z_5 = \{10 + 15i; 1\}$$

for the initial disks containing the zeros of g, i.e., the eigenvalues of H. The maximal radii $r^{(m)} = \max \operatorname{rad} Z_i^{(m)}$ $(m = 1, 2)$ of the produced disks are displayed in Table 5.7.

From Table 5.7, we observe that the applied inclusion methods converge very fast. The explanation for this extremely rapid convergence lies in the fact that the eigenvalues of Hessenberg's matrix considered in this example are very close to the diagonal elements. Because of the closeness to the desired zeros, the centers of initial disks cause very fast convergence of the sequences of inclusion disk centers, which provides fast convergence of the sequences of radii.

Table 5.7 The enclosure of the eigenvalues of Hessenberg's matrix

Methods	$r^{(1)}$	$r^{(2)}$
(ITS) $\alpha = 1$	2.73(-10)	4.92(-43)
(ITS) $\alpha = 1/2$	2.39(-10)	3.65(-43)
(ITS) $\alpha = 1/(n-1)$	2.21(-10)	3.02(-43)
(ITS) $\alpha = 0$	2.04(-10)	2.38(-43)
(ITS) $\alpha = -1$	2.73(-10)	2.74(-43)
(5.75)	5.64(-7)	1.71(-37)
(5.76)	3.27(-7)	1.60(-28)

We note that, as in Example 5.4, the Weierstrass-like method (5.77) diverged.

In Sect. 5.3, we presented a new one-parameter family of iterative methods for the simultaneous inclusion of simple complex zeros of a polynomial. According to a hundred numerical examples and the theoretical convergence analysis given in [125]–[127], the characteristics and advantages of this family can be summarized as follows:

- The produced enclosing disks enable automatic determination of rigorous error bounds of the obtained approximations.
- The proposed family is of general type and includes previously derived methods of the square root type.
- Numerical examples demonstrate stable and fast convergence of the family (ITS); furthermore, the methods of this family compete with the existing inclusion methods of the fourth order (5.75) and (5.76), they even produce tighter disks in some cases; moreover, numerical experiments show that a variation of the parameter α can often provide a better approach to the wanted zeros compared with (5.75) and (5.76). See Examples 5.4 and 5.5.
- The quadratically convergent Weierstrass-like method (5.77) diverged not only in the displayed examples, but also in solving numerous polynomial equations. This means that the application of quadratically convergent methods, such as the Weierstrass-like method (5.77), is not always better than higher-order methods. Such outcome partly arises from the fact that the product of disks is not an exact operation in circular interval arithmetic (see (1.66)) and usually gives enlarged disks. In this manner, the disk in the denominator of (5.77) can include the number 0 and the method becomes undefined. This disadvantage of the inclusion method (5.77) can be avoided if we use the following iterative formula

$$\hat{Z}_i = z_i - P(z_i) \prod_{\substack{j=1 \\ j \neq i}}^{n} (z_i - Z_j)^{-1} \quad (i \in \boldsymbol{I}_n) \tag{5.78}$$

instead of (5.77). In this way, disks containing 0 are eliminated, but the product of inverse disks still generates large disks and the iterative process (5.78) will break in the next iterations.

- The convergence order of the proposed family is 4; it can be significantly increased by suitable (already calculated) corrections with negligible number of additional operations, providing in this way a high computational efficiency, see [127].
- A slight modification of the fixed point relation, which served as the base for the construction of the considered algorithm, can provide the simultaneous inclusion of multiple zeros, see [125].

In our experiments, we used various values of the parameter α belonging to the disk $\{0; 3\}$. We could not find a specific value of α giving a method from the family (ITS) which would be asymptotically best for all P. All tested methods presented similar behavior for the values of $\alpha \in \{0; 3\}$ and very fast convergence for good initial approximations. A number of numerical examples showed only that one of the tested methods is the best for some polynomials, while the other one is the best for other polynomials. Actually, the convergence behavior strongly depends on initial approximations, their distribution, and the structure of tested polynomials.

References

1. Aberth, O.: Iteration methods for finding all zeros of a polynomial simultaneously. Math. Comp., **27**, 339–344 (1973)
2. Alefeld, G., Herzberger, J.: On the convergence speed of some algorithms for the simultaneous approximation of polynomial zeros. SIAM J. Numer. Anal., **11**, 237–243 (1974)
3. Alefeld, G., Herzberger, J.: Introduction to Interval Computation. Academic Press, New York (1983)
4. Bateman, H.: Halley's methods for solving equations. Amer. Math. Monthly, **45**, 11–17 (1938)
5. Batra, P.: Improvement of a convergence condition for the Durand-Kerner iteration. J. Comput. Appl. Math., **96**, 117–125 (1998)
6. Bell, E.T.: Exponential polynomials. Math. Ann., **35**, 258–277 (1934)
7. Bini, D.A., Fiorentino, G.: Design, analysis and implementation of a multiprecision polynomial rootfinder. Numer. Algor., **23**, 127–173 (2000)
8. Bodewig, E.: Sur la méthode Laguerre pour l'approximation des racines de certaines équations algébriques et sur la critique d'Hermite. Indag. Math., **8**, 570–580 (1946)
9. Börsch-Supan, W.: A posteriori error bounds for the zeros of polynomials. Numer. Math., **5**, 380–398 (1963)
10. Börsch-Supan, W.: Residuenabschätzung für Polynom-Nullstellen mittels Lagrange-Interpolation. Numer. Math., **14**, 287–296 (1970)
11. Braess, D., Hadeler K.P.: Simultaneous inclusion of the zeros of a polynomial. Numer. Math., **21**, 161–165 (1973)
12. Brugnano, L., Trigiante, D.: Polynomial roots: the ultimate answer? Linear Algebra Appl., **225**, 207–219 (1995)
13. Carstensen, C.: Anwendungen von Begleitmatrizen. Z. Angew. Math. Mech., **71**, 809–812 (1991)
14. Carstensen, C.: Inclusion of the roots of a polynomial based on Gerschgorin's theorem. Numer. Math., **59**, 349–360 (1991)
15. Carstensen, C.: On quadratic-like convergence of the means for two methods for simultaneous root finding of polynomials. BIT, **33**, 64–73 (1993)
16. Carstensen, C., Petković, M.S.: On iteration methods without derivatives for the simultaneous determination of polynomial zeros. J. Comput. Appl. Math., **45**, 251–266 (1993)
17. Carstensen, C., Petković, M.S.: An improvement of Gargantini's simultaneous inclusion method for polynomial roots by Schroeder's correction. Appl. Numer. Math., **13**, 453–468 (1994)
18. Carstensen, C., Sakurai, T.: Simultaneous factorization of a polynomial by rational approximation. J. Comput. Appl. Math., **61**, 165–178 (1995)

19. Chen, P.: Approximate zeros of quadratically convergent algorithms. Math. Comp., **63**, 247–270 (1994)
20. Cira, O.: Metode Numerice Pentru Rezolvarea Ecuaţiilor Algebrice (in Rumanian). Editura Academiei Romane, Bucureşti (2005)
21. Collatz, L.: Functional Analysis and Numerical Mathematics. Academic Press, New York-San Francisco-London (1966)
22. Cosnard, M., Fraigniaud, P.: Asynchronous Durand-Kerner and Aberth polynomial root finding methods on a distributed memory multicomputer. Parallel Computing, **9**, 79–84 (1989)
23. Cosnard, M., Fraigniaud, P.: Finding the roots of a polynomial on an MIMD multicomputer. Parallel Computing, **15**, 75–85 (1990)
24. Cosnard, M., Fraigniaud, P.: Analysis of asynchronous polynomial root finding methods on a distributed memory multicomputer. IEEE Trans. on Parallel and Distributed Systems, **5**, 639–648 (1994)
25. Curry, J.H.: On zero finding methods of higher order from data at one point. J. Complexity, **5**, 219–237 (1989)
26. Curry, J.H., Van Vleck, E.S.: On the theory and computation of approximate zeros. Preprint.
27. Davies, M., Dawson, B.: On the global convergence of Halley's iteration formula. Numer. Math., **24**, 133–135 (1975)
28. Dekker, T.J.: Newton-Laguerre iteration. Programation en Mathématique Numériques, Colloq. Internat. du CNRS, Besançon (1968), pp. 189–200
29. Dieudonné, J.: Foundations of Modern Analysis. Academic Press, New York (1960)
30. Dochev, K.: Modified Newton method for the simultaneous approximate calculation of all roots of a given algebraic equation (in Bulgarian). Mat. Spis. B"lgar. Akad. Nauk, **5**, 136–139 (1962)
31. Dochev, K., Byrnev, P.: Certain modifications of Newton's method for the approximate solution of algebraic equations. Ž. Vyčisl. Mat. i Mat. Fiz., **4**, 915–920 (1964)
32. Durand, E.: Solution numériques des équations algébriques, Tom. I: Équations du Type F(x) = 0; Racines d'un Polynôme. Masson, Paris (1960)
33. Ehrlich, L.W.: A modified Newton method for polynomials. Comm. ACM, **10**, 107–108 (1967)
34. Ellis, G.H., Watson, L.T.: A parallel algorithm for simple roots of polynomials. Comput. Math. Appl., **2**, 107–121 (1984)
35. Elsner, L.: Remark on simultaneous inclusion of the zeros of a polynomial by Gerschgorin's theorem. Numer. Math., **21**, 425–427 (1973)
36. Falconer, K.: Fractal Geometry. Mathematical Foundation and Applications, John Wiley and Sons (1990)
37. Farmer, M.R., Loizou, G.: A class of iteration functions for improving, simultaneously, approximations to the zeros of a polynomial. BIT, **15**, 250–258 (1975)
38. Farmer, M.R., Loizou, G.: An algorithm for total, or parallel, factorization of a polynomial. Math. Proc. Cambridge Philos. Soc., **82**, 427–437 (1977)
39. Farmer, M.R., Loizou, G.: Locating multiple zeros interactively. Comput. Math. Appl., **11**, 595–603 (1985)
40. Fiedler, M.: Expressing a polynomial as the characteristic polynomial of a symmetric matrices. Linear Algebra Appl., **141**, 265–270 (1990)
41. Ford, W.F., Pennline, J.A.: Accelerated convergence in Newton's method. SIAM Rev., **38**, 658–659 (1996)
42. Fourier, J.B.J.: Oeuvres de Fourier, Vol. II. Gauthier-Villars, Paris, 249–250 (1890)
43. Fraigniaud, P.: The Durand-Kerner polynomial root finding method in case of multiple roots. BIT, **31**, 112–123 (1991)
44. Freeman, T.L.: Calculating polynomial zeros on a local memory parallel computer. Parallel Computing, **12**, 351–358 (1989)
45. Gander, W.: On Halley's iteration methods. Amer. Math. Monthly, **92**, 131–134 (1985)

46. Gargantini, P.: Parallel algorithms for the determination of polynomial zeros. In: Thomas, R., Williams, H.C. (eds) Proc. III Manitoba Conf. on Numer. Math., Utilitas Mathematica Publ. Inc., Winnipeg (1974), pp. 195–211

47. Gargantini, I.: Parallel Laguerre iterations: The complex case. Numer. Math., **26**, 317–323 (1976)

48. Gargantini, I.: Further application of circular arithmetic: Schroeder-like algorithms with error bound for finding zeros of polynomials. SIAM J. Numer. Anal., **15**, 497–510 (1978)

49. Gargantini, I.: Parallel square-root iterations for multiple roots. Comput. Math. Appl., **6**, 279–288 (1980)

50. Gargantini, I., Henrici, P.: Circular arithmetic and the determination of polynomial zeros. Numer. Math., **18**, 305–320 (1972)

51. Gerlach, J.: Accelerated convergence in Newton's method. SIAM Rev., **36**, 272–276 (1994)

52. Green, M.W., Korsak, A.J., Pease, M.C.: Simultaneous iteration toward all roots of a complex polynomial. SIAM Rev., **18**, 501–502 (1976)

53. Gregg, W., Tapia, R.: Optimal error bounds for the Newton-Kantorowich theorem. SIAM J. Numer. Anal., **11**, 10–13 (1974)

54. Halley, E.: A new, exact, and easy method of findinng the roots of any equations generally, and that without any previous reduction. ($\mathcal{ABRIDGED}$, by C. Hutton, G. Shaw, R. Pearson, translated from Latin), Phil. Trans. Roy. Soc. London III (1809)

55. Hansen, E., Patrick, M.: A family of root finding methods. Numer. Math., **27**, 257–269 (1977)

56. Hansen, E., Patrick, M., Rusnak, J.: Some modifications of Laguerre's method. BIT, **17**, 409–417 (1977)

57. Henrici, P.: Applied and Computational Complex Analysis, Vol. I. John Wiley and Sons, New York (1974)

58. Herceg, Ð.D.: Computer implementation and interpretation of iterative methods for solving equations. Master Thesis, University of Novi Sad, Novi Sad (1997)

59. Herceg, Ð.D.: An algorithm for localization of polynomial zeros. In: Tošić, R. Budimac, Z. (eds) Proc. on VIII Conf on Logic and Computer Science, Institute of Mathematics, Novi Sad (1997), pp. 67–75

60. Herceg, Ð.D.: Convergence of simultaneous methods for finding polynomial zeros. Ph. D. Thesis, University of Novi Sad, Novi Sad (1999)

61. Hribernig, V., Stetter, H.J.: Detection and validation of clusters of polynomial zeros. J. Symbolic Comput., **24**, 667–681 (1997)

62. Huang, Z., Zhang, S.: On a family of parallel root-finding methods for a generalized polynomials. Appl. Math. Comput., **91**, 221–231 (1998)

63. Igarashi, M., Nagasaka, H.: Relationships between the iteration times and the convergence order for Newton-Raphson like methods (in Japanese). J. Inform. Process., **22**, 1349–1354 (1991)

64. Ilić, S., Petković, M.S., Herceg, Ð.: A note on Babylonian square-root algorithm and related variants. Novi Sad J. Math., **26**, 155–162 (1996)

65. Iokimidis, I.O., Anastasselou, E.G.: On the simultaneous determination of zeros of analytic or sectionally analytic functions. Computing, **36**, 239–246 (1986)

66. Jamieson, L.H., Rice, T.A.: A highly parallel algorithms for root extraction. IEEE Trans. on Comp., **28**, 443–449 (1989)

67. Jenkins, M.A., Traub, J.F.: A three stage variable-shift iteration for polynomial zeros. SIAM J. Numer. Anal., **7**, 545–566 (1970)

68. Jovanović, B.: A method for obtaining iterative formulas of higher order. Mat. Vesnik, **9**, 365–369 (1972)

69. Kanno, S., Kjurkchiev, N., Yamamoto, T.: On some methods for the simultaneous determination of polynomial zeros. Japan J. Indust. Appl. Math., **2**, 267–288 (1996)

70. Kantorovich, L.V.: Functional analysis and applied mathematics (in Russian). Uspekhi Mat. Nauk., **3**, 89–185 (1948)

71. Kantorovich, L.V., Akilov, G.: Functional Analysis in Normed Spaces. MacMillan, New York (1964)

72. Kerner, I.O.: Ein Gesamtschrittverfahren zur Berechnung der Nullstellen von Polynomen. Numer. Math., **8**, 290–294 (1966)

73. Kim, M.: Computational complexity of the Euler type algorithms for the roots of complex polynomials. Ph. D. Thesis. City University of New York, New York (1985)

74. Kim, M.: On approximate zeros and root finding algorithms for a complex polynomial. Math. Comp., **51**, 707–719 (1988)

75. King, R.F.: Improving the Van de Vel root-finding method. Computing, **30**, 373–378 (1983)

76. Knuth, D.: The Art of Programming, Vol. 2. Addison-Wesley, New York (1969)

77. Kravanja, P.: A modification of Newton's method for analytic mappings having multiple zeros. Computing, **62**, 129–145 (1999)

78. Kravanja, P.: On computing zeros of analytic functions and related problems in structured numerical linear algebra. Ph. D. Thesis, Katholieke Universiteit Leuven, Lueven (1999)

79. Kravanja, P., Van Barel, M.: Computing the Zeros of Analytic Functions. Lecture Notes in Mathematics 1727, Springer-Verlag, Berlin (2000)

80. Kravanja, P., Sakurai, T., Van Barel, M.: On locating clusters of zeros of analytic functions. BIT, **39**, 646–682 (1999)

81. Kyurkchiev, N.: Initial Approximations and Root Finding Methods. Wiley-VCH, Berlin (1998)

82. Kyurkchiev, N., Markov, S.: Two interval methods for algebraic equations with real roots. Pliska, **5**, 118–131 (1983)

83. Lang, S.: Real Analysis. Addison-Wesley, Reading, Massachusetts (1983)

84. Li, T.Y., Zeng, Z.: The Laguerre iteration in solving the symmetric tridiagonal eigenproblem. Revisited, SIAM J. Sci. Comput., **5**, 1145–1173 (1994)

85. Maehly, V.H.: Zur iterativen Auflösung algebraischer Gleichungen. Z. Angew. Math. Phys., **5**, 260–263 (1954)

86. Malek, F., Vaillancourt, R.: Polynomial zero finding iterative matrix algorithm. Comput. Math. Appl., **29**, 1–13 (1995)

87. Marden, M.: The Geometry of Polynomials. Mathematical surveys, Amer. Math. Soc., Providence, Rhode Island (1966)

88. Milovanović, G.V.: A method to accelerate iterative processes in Banach space. Univ. Beograd. Publ. Elektrotehn. Fak. Ser. Mat. Fiz., **470**, 67–71 (1974)

89. Miyakoda, T.: Balanced convergence of iterative methods to a multiple zero of a complex polynomial. J. Comput. Appl. Math., **39**, 201–212 (1992)

90. Neumaier, A.: An existence test for root clusters and multiple roots. ZAMM, **68**, 256–257 (1988)

91. Neumaier, A.: Enclosing clusters of zeros of polynomials. J. Comput. Appl. Math., **156**, 389–401 (2003)

92. Nicolas, J.L., Schinzel, A.: Localisation des zéros de polynomes intervenant end théorie du signal. Research report, University of Lyon 1, Lyon (1988)

93. Niu, X.M., Sakurai, T.: A method for finding the zeros of polynomials using a companion matrix. Japan J. Idustr. Appl. Math., **20**, 239–256 (2003)

94. Niu, X.M., Sakurai, T., Sugiura, H.: A verified method for bounding clusters of zeros of analytic functions, J. Comput. Appl. Math., **199**, 263–270 (2007)

95. Nourein, A.W.M.: An iteration formula for the simultaneous determination of the zeroes of a polynomial. J. Comput. Appl. Math., **4**, 251–254 (1975)

96. Nourein, A.W.M.: An improvement on Nourein's method for the simultaneous determination of the zeroes of a polynomial (an algorithm). J. Comput. Appl. Math., **3**, 109–110 (1977)

97. Nourein, A.W.M.: An improvement on two iteration methods for simultaneous determination of the zeros of a polynomial. Internat. J. Comput. Math., **6**, 241–252 (1977)

98. Ortega, J.M., Rheinboldt, W.C.: Iterative Solution of Nonlinear Equations in Several Variables. Academic Press, New York (1970)
99. Ostrowski, A.M.: Solution of Equations in Euclidian and Banach Spaces. Academic Press, New York (1973)
100. Pan, V.Y.: Optimal and nearly optimal algorithms for approximating polynomial zeros. Comput. Math. Appl., **31**, 97–138 (1996)
101. Pan, V.Y.: Solving a polynomial equation: some history and recent progress. SIAM Rev., **39**, 187–220 (1997)
102. Parlett, B.: Laguerre's method applied to the matrix eigenvalue problem. Math. Comp., **18**, 464–485 (1964)
103. Pasquini, L., Trigiante, D.: A globally convergent method for simultaneously finding polynomial roots. Math. Comp., **44**, 135–149 (1985)
104. Peitgen, H.-O., Richter, P.H.: The Beauty of Fractals. Springer-Verlag, Berlin (1986)
105. Petković, L.D., Petković, M.S.: The representation of complex circular functions using Taylor series. Z. Angew. Math. Mech., **61**, 661–662 (1981).
106. Petković, M.S.: On a generalization of the root iterations for polynomial complex zeros in circular interval arithmetic. Computing, **27**, 37–55 (1981)
107. Petković, M.S.: On an iteration method for simultaneous inclusion of polynomial complex zeros. J. Comput. Appl. Math., **8**, 51–56 (1982)
108. Petković, M.S.: On Halley-like algorithms for the simultaneous approximation of polynomial complex zeros. SIAM J. Numer. Math., **26**, 740–763 (1989)
109. Petković, M.S.: Iterative Methods for Simultaneous Inclusion of Polynomial Zeros. Springer-Verlag, Berlin (1989)
110. Petković, M.S.: On initial conditions for the convergence of simultaneous root finding methods. Computing, **57**, 163–177 (1996)
111. Petković, M.S., Carstensen, C.: On some improved inclusion methods for polynomial roots with Weierstrass' corrections. Comput. Math. Appl., **25**, 59–67 (1993)
112. Petković, M.S., Carstensen, C., Trajković, M.: Weierstrass' formula and zero-finding methods. Numer. Math., **69**, 353–372 (1995)
113. Petković, M.S., Herceg, D.: On rediscovered iteration methods for solving equations. J. Comput. Appl. Math., **107**, 275–284 (1999)
114. Petković, M.S., Herceg, Đ.: Point estimation and safe convergence of root-finding simultaneous methods. Sci. Rev., **21/22**, 117–130 (1996)
115. Petković, M.S., Herceg, Đ.: Börsch-Supan-like methods: Point estimation and parallel implementation. Intern. J. Comput. Math., **64**, 117–130 (1997)
116. Petković, M.S., Herceg, Đ.: On the convergence of Wang-Zheng's method. J. Comput. Appl. Math., **91**, 123–135 (1998)
117. Petković, M.S., Herceg, Đ.: Point estimation of simultaneous methods for solving polynomial equations: a survey. J. Comput. Appl. Math., **136**, 183–207 (2001)
118. Petković, M.S., Herceg, Đ., Ilić, S.: Point Estimation Theory and its Applications. Institute of Mathematics, Novi Sad, (1997)
119. Petković, M.S., Herceg, Đ., Ilić, S.: Point estimation and some applications to iterative methods. BIT, **38**, 112–126 (1998)
120. Petković, M.S., Herceg, Đ., Ilić, S.: Safe convergence of simultaneous methods for polynomial zeros. Numer. Algor., **17**, 313–331 (1998)
121. Petković, M.S., Ilić, S.: Point estimation and the convergence of the Ehrlich-Aberth method. Publ. Inst. Math., **62**, 141–149 (1997)
122. Petković, M.S., Ilić, S., Petković, I.: A posteriori error bound methods for the inclusion of polynomial zeros. J. Comput. Appl. Math., **208**, 316–330 (2007)
123. Petković, M.S., Ilić, S., Rančić, L.: The convergence of a family of parallel zero-finding methods. Comput. Math. Appl., **48**, 455–467 (2004)
124. Petković, M.S., Ilić, S., Tričković, S.: A family of simultaneous zero finding methods. Comput. Math. Appl., **34**, 49–59 (1997)
125. Petković, M.S., Milošević, D.: A higher order family for the simultaneous inclusion of multiple zeros of polynomials. Numer. Algor., **39**, 415–435 (2005)

126. Petković, M.S., Milošević, D.M.: A new higher-order family of inclusion zero-finding methods. J. Comp. Appl. Math., **182**, 416–432 (2005)
127. Petković, M.S., Milošević, D.M.: On a new family of simultaneous methods with corrections for the inclusion of polynomial zeros. Intern. J. Comput. Math., **83**, 299–317 (2006)
128. Petković, M.S., Petković, L.D.: On a computational test for the existence of polynomial zero. Comput. Math. Appl., **17**, 1109–1114 (1989)
129. Petković, M.S., Petković, L.D.: Complex Interval Arithmetic and its Applications. Wiley-VCH, Berlin-Weinhein-New York, (1998)
130. Petković, M.S., Petković, L.D.: On the convergence of the sequences of Gerschgorin-like disks. Numer. Algor., **42**, 363–377 (2006)
131. Petković, M.S., Petković, L.D.: On a cubically convergent derivative free root finding method. Intern. J. Comput. Math., **84**, 505–513 (2007)
132. Petković, M.S., Petković, L.D., Herceg, Đ.: Point estimation of a family of simultaneous zero-finding methods. Comput. Math. Appl., **36**, 1–12 (1998)
133. Petković, M.S., Petković, L.D., Ilić, S.: On the guaranteed convergence of Laguerre-like method. Comput. Math. Appl., **46**, 239–251 (2003)
134. Petković, M.S., Petković, L.D., Živković, D.: Laguerre-like methods for the simultaneous approximation of polynomial zeros. In: Alefeld, G., Chen, X. (eds) Topics in Numerical Analysis with Special Emphasis on Nonlinear Problems, Springer-Verlag, New York-Wien (2001), pp. 189–210
135. Petković, M.S., Petković, L.D., Živković, D.: Hansen-Patrick's family is of Laguerre's type. Novi Sad J. Math., **30**, 109–115 (2003)
136. Petković, M.S., Rančić, L.: On the guaranteed convergence of the square-root iteration method. J. Comput. Appl. Math., **170**, 169–179 (2004)
137. Petković, M.S., Rančić, L.: On the guaranteed convergence of the fourth order simultaneous method for polynomial zeros. Appl. Math. Comput., **155**, 531–543 (2004)
138. Petković, M.S., Rančić, L.: A new-fourth order family of simultaneous methods for finding polynomial zeros. Appl. Math. Comp., **164**, 227–248 (2005)
139. Petković, M.S., Rančić, L.: On a family of root-finding methods with accelerated convergence. Comput. Math. Appl., **51**, 999–1010 (2006)
140. Petković, M.S., Rančić, L., Petković, L.D.: Point estimation of simultaneous methods for solving polynomial equations: a survey (II). J. Comput. Appl. Math., **205**, 32–52 (2007)
141. Petković, M.S., Sakurai, T., Rančić, L.: Family of simultaneous methods of Hansen-Patrick's type. Appl. Numer. Math., **50**, 489–510 (2004)
142. Petković, M.S., Stefanović, L.V.: On some improvements of square root iteration for polynomial complex zeros. J. Comput. Appl. Math., **15**, 13–25 (1986)
143. Petković, M.S., Stefanović, L.V.: On some parallel higher-order methods of Halley's type for finding multiple polynomial zeros. In: Milovanović, G.V. (ed) Numerical Methods and Approximation Theory, Faculty of Electronic Engineering, Niš (1988), pp. 329–337
144. Petković, M.S., Stefanović, L.V., Marjanović, Z.M.: On the R-order of some accelerated methods for the simultaneous finding polynomial zeros. Computing, **49**, 349–361 (1993)
145. Petković, M.S., Tričković, S.: Tchebychev-like method for simultaneous finding zeros of analytic functions. Comput. Math. Appl., **31**, 85–93 (1996)
146. Petković, M.S., Tričković, S., Herceg, Đ.: On Euler-like methods for the simultaneous approximation of polynomial zeros. Japan J. Indust. Appl. Math., **15**, 295–315 (1998)
147. Prešić, M.: An iterative procedure for determination of k roots of a polynomial (in Serbian). Ph. D. Thesis, University of Belgrade, Belgrade (1971)
148. Prešić, M.: A convergence theorem for a method for simultaneous determination of all zeros of a polynomial. Publications de l'Institut Mathématique, **28**, 159–168 (1980)
149. Prešić, S.B.: Un procédé itératif pour la factorisation des polynômes. C. R. Acad. Sci. Paris Sér. A, **262**, 862–863 (1966)

150. Proinov, P.: A new semilocal convergence theorem for the Weierstrass method from data at one point. Compt. Rend. Acad. Bulg. Sci., **59**, 131–136 (2006)
151. Proinov, P.: Semilocal convergence of two iterative methods for simultaneous computation of polynomial zeros. Compt. Rend. Acad. Bulg. Sci., **59**, 705–712 (2006)
152. Rall, L.: A note on the convergence of Newton's method. SIAM J. Numer. Anal., **11**, 34–36 (1974)
153. Rančić, L.Z.: Simultaneous methods for solving algebraic equations. Ph. D. Thesis, University of Niš, Niš (2005)
154. Riordan, J.: Combinatorial Identities. John Wiley and Sons, New York-London-Sydney (1968)
155. Rump, S.M.: Ten methods to bound multiple roots of polynomials. J. Comput. Appl. Math., **156**, 403–432 (2003)
156. Sakurai, T., Petković, M.S.: On some simultaneous methods based on Weierstrass-Dochev correction. J. Comput. Appl. Math., **72**, 275–291 (1996)
157. Sakurai, T., Torii, T., Sugiura, H.: A high-order iterative formula for simultaneous determination of zeros of a polynomial. J. Comput. Appl. Math., **38**, 387–397 (1991)
158. Sakurai, T., Sugiura, H., Torii, T.: Numerical factorization of a polynomial by rational Hermite interpolation. Numer. Algor., **3**, 411–418 (1992)
159. Schmeisser, G.: A real symmetric tridiagonal matrix with a given characteristic polynomial. Linear Algebra Appl., **193**, 11–18 (1993)
160. Schröder, E.: Über unendlich viele Algorithmen zur Auflösung der Gleichungen. Math. Ann., **2**, 317–365 (1870)
161. Sendov, B., Andreev, A., Kyurkchiev, N.: Numerical Solution of Polynomial Equations (Handbook of Numerical Analysis). Vol. VIII, Elsevier Science, New York (1994)
162. Sendov, B., Popov, V.: Numerical Methods, Part I (in Bulgarian). Nauka i Izkustvo, Sofia (1976)
163. Shub, M., Smale, S.: Computational complexity: on the geometry of polynomials and a theory of costs, I. Ann. Sci. École Norm. Sup., **18**, 107–142 (1985)
164. Shub, M., Smale, S.: Computational complexity: on the geometry of polynomials and a theory of costs, II. SIAM J. Comput., **15**, 145–161 (1986)
165. Smale, S.: The fundamental theorem of algebra and complexity theory. Bull. Amer. Math. Soc., **4**, 1–35 (1981)
166. Smale, S.: On the efficiency of algorithms of analysis. Bull. Amer. Soc. (N.S.), **13**, 87–121 (1985)
167. Smale, S.: Newton's method estimates from data at one point. In: Ewing, R.E, Gross, K.I., Martin, C.F. (eds) The Merging Disciplines: New Directions in Pure, Applied and Computational Mathematics, Springer-Verlag, New York (1986), pp. 185–196
168. Smith, B.T.: Error bounds for zeros of a polynomial based upon Gerschgorin's theorem. J. Assoc. Comput. Mach., **17**, 661–674 (1970)
169. Steffensen, I.F.: Remark on iteration. Skand. Aktuarietidskr., **16**, 64–72 (1933)
170. Stoer, J., Bulirsch, R.: Einführung in die Numerische Mathematik II. Springer-Verlag, Berlin (1973)
171. Tanabe, K.: Behavior of the sequences around multiple zeros generated by some simultaneous methods for solving algebraic equations (in Japanese). Tech. Rep. Inf. Procces. Numer. Anal., **4-2**, 1–6 (1983)
172. Traub, J.F.: Iterative Methods for the Solution of Equations. Prentice Hall, New Jersey (1964)
173. Traub, J.F., Wozniakowski, H.: Convergence and complexity of Newton iteration for operator equations. J. Assoc. Comp. Mach., **29**, 250–258 (1979)
174. Tričković, S.: Iterative methods for finding polynomial zeros (in Serbian). Ph. D. Thesis, University of Novi Sad, Novi Sad (1997)
175. Van der Sluis, A.: Upper bounds for roots of polynomials. Numer. Math., **15**, 250–262 (1970)
176. Varjuhin, V.A., Kasjanjuk, S.A.: On iterative methods for refinement of roots of equations (in Russian). Ž. Vyčisl. Mat. i Mat. Fiz., **9**, 684–687 (1969)

177. Wang, D., Wu, Y.: Some modifications of the parallel Halley iteration method and their convergence. Computing, **38**, 75–87 (1987)

178. Wang, D., Zhao, F.: The theory of Smale's point estimation and its application. J. Comput. Appl. Math., **60**, 253–269 (1995)

179. Wang, X.: Journal of Hangzhou University (Natural Science) (Chinese), **3**, 63–70 (1966)

180. Wang, X.: Convergence of Newton's method and inverse function theorem. Math. Comp., **68**, 169–186 (1999)

181. Wang, X., Han, D.: On dominating sequence method in the point estimate and Smale's theorem. Scientia Sinica Ser. A, 905–913 (1989)

182. Wang, X., Zheng, S.: A family of parallel and interval iterations for finding all roots of a polynomial simultaneously with rapid convergence (I). J. Comput. Math., **1**, 70–76 (1984)

183. Wang, X., Zheng, S.: The quasi-Newton method in parallel circular iteration. J. Comput. Math., **4**, 305–309 (1984)

184. Wang, X., Zheng, S.: Parallel Halley iteration method with circular arithmetic for finding all zeros of a polynomial. Numer. Math., J. Chenese Univ., **4**, 308–314 (1985)

185. Wang, X., Zheng, S.: A family of parallel and interval iterations for finding all roots of a polynomial simultaneously with rapid convergence (II) (in Chinese). J. Comput. Math., **4**, 433–444 (1985)

186. Wang, X., Zheng, S., Shen, G.: Bell's disk polynomial and parallel disk iteration. Numer. Math., J. Chinese Univ., **4**, 328–345 (1987)

187. Weierstrass, K.: Neuer Beweis des Satzes, dass jede ganze rationale Funktion einer Veränderlichen dargestellt werden kann als ein Product aus linearen Funktionen derselben Veränderlichen. Ges. Werke, **3**, 251–269 (1903) (Johnson Reprint Corp., New York 1967)

188. Werner, W.: On the simultaneous determination of polynomial roots. In: Iterative Solution of Nonlinear Systems of Equations, Lecture Notes in Mathematics **953**, Springer-Verlag, Berlin (1982), pp. 188–202

189. Wilf, H.S.: A global bisection algorithm for computing the zeros of polynomials in the complex plane. J. Assoc. Comput. Mach., **25**, 415–420 (1978)

190. Wilkinson, J.H.: Rounding Errors in Algebraic Processes. Prentice Hall, New Jersey (1963)

191. Yakoubsohn, J.C.: Finding a cluster of zeros of univariate polynomials. J. Complexity, **16**, 603–638 (2000)

192. Yamagishi, Y.: Master Thesis. Kyoto University, Kyoto (1991)

193. Yamamoto, T., Furukane, U., Nogura, K.: The Durand-Kerner and the Aberth methods for solving algebraic equations (in Japanese). Jyo-ho Syori (Information Processing), **18**, 566–571 (1977)

194. Yamamoto, T., Kanno, S., Atanassova, L., Validated computation of polynomial zeros by the Durand-Kerner method. In: Herzberger, J. (ed) Topics in Validated Computations, Elsevier Science, B.V., Amsterdam (1994)

195. Zhao, F., Wang, D.: The theory of Smale's point estimation and the convergence of Durand-Kerner program (in Chinese). Math. Numer. Sinica, **15**, 196–206 (1993)

196. Zheng, S., Sun, F.: Some simultaneous iterations for finding all zeros of a polynomial with high order of convergence. Appl. Math. Comput., **99**, 233–240 (1999)

Glossary

$\mathbb{C}$ the set of complex numbers, 32

$K(\mathbb{C})$ the set of circular disks, 32

$Z = \{c; r\}$ circular disk $= \{z \in \mathbb{C} : |z - c| \leq r\} \in K(\mathbb{C})$, 32

mid Z center of a disk Z, 32

rad Z radius of a disk Z, 32

Z^{-1} exact inverse of a disk Z, 32

Z^{I_c} centered inverse of a disk Z, 32

$Z_1 \cap Z_2$ intersection of two disks, 33

$Z_1 \subseteq Z_2$ inclusion of disks, 33

$Z^{1/2}$ square root of a disk Z, 34

I_n index set, $I_n = \{1, \ldots, n\}$, 2

P monic algebraic polynomial, $P(z) = z^n + a_{n-1}z^{n-1} + \cdots + a_1 z + a_0$, 2

z_i approximation to the zero ζ_i of a polynomial P, 2

$\hat{z}_i$ the next approximation to the zero ζ_i, 4

$\delta_{k,i} = P^{(k)}(z_i)/P(z_i)$ $(k = 1, 2, \ldots)$, 5

$\Delta_i = \left(P'(z_i)^2 - P(z_i)P''(z_i)\right)/P(z_i)^2$, 163

W_i Weierstrass' correction $= P(z_i)/\prod_{j \neq i}(z_i - z_j)$, 2

N_i Newton's correction $= P(z_i)/P'(z_i)$, 8

H_i Halley's correction $= \left(P'(z_i)/P(z_i) - P''(z_i)/(2P'(z_i))\right)^{-1}$, 6

w maximal Weierstrass' correction $= \max_{1 \leq i \leq n} |W_i|$, 68

f function whose zero ζ is sought, 35

f^{-1} the inverse function to f, 38

$f^{(k)}$ the kth derivative of a complex function f, 46

$f^{(k)}_{(z)}$ the kth Fréchet derivative at the point z, 37

ζ a zero of f, 35

ζ_i a zero of a polynomial P, 2

μ the multiplicity of the zero ζ of a function f, 21

μ_i the multiplicity of the zero ζ_i of a polynomial P, 21

$E_k(z, h, f)$ the kth incremental Euler algorithm, 49

d minimal distance between approximations $= \min\limits_{j \neq i} |z_i - z_j|$, 61

c_n i-factor, $w \leq c_n d$, 69

$z_i^{(m)}$ approximation to the zero ζ_i in the mth iteration, 71

$u_i^{(m)}$ error in the mth iteration $= z_i^{(m)} - \zeta_i$, 97

γ_n convergence factor, 97

$C_i^{(m)}$ iterative correction in the mth iteration, 71

β_n contraction factor, $|C_i^{(m+1)}| < \beta_n |C_i^{(m)}|$, 74

$\mathcal{O}$ same order of magnitude (of real numbers), 23

$\mathcal{O}_M$ same order of magnitude (of complex numbers), 135

O_R R-order of convergence, 181

$\rho^{(m)}$ a measure of the separation of inclusion disks generated in the mth iteration, 190

$\approx$ approximate equality between numbers, 7

Index

Lecture Notes in Mathematics

For information about earlier volumes
please contact your bookseller or Springer
LNM Online archive: springerlink.com

Vol. 1911: A. Bressan, D. Serre, M. Williams, K. Zumbrun, Hyperbolic Systems of Balance Laws. Cetraro, Italy 2003. Editor: P. Marcati (2007)

Vol. 1912: V. Berinde, Iterative Approximation of Fixed Points (2007)

Vol. 1913: J.E. Marsden, G. Misiołek, J.-P. Ortega, M. Perlmutter, T.S. Ratiu, Hamiltonian Reduction by Stages (2007)

Vol. 1914: G. Kutyniok, Affine Density in Wavelet Analysis (2007)

Vol. 1915: T. Bıyıkoğlu, J. Leydold, P.F. Stadler, Laplacian Eigenvectors of Graphs. Perron-Frobenius and Faber-Krahn Type Theorems (2007)

Vol. 1916: C. Villani, F. Rezakhanlou, Entropy Methods for the Boltzmann Equation. Editors: F. Golse, S. Olla (2008)

Vol. 1917: I. Veselić, Existence and Regularity Properties of the Integrated Density of States of Random Schrödinger (2008)

Vol. 1918: B. Roberts, R. Schmidt, Local Newforms for GSp(4) (2007)

Vol. 1919: R.A. Carmona, I. Ekeland, A. Kohatsu-Higa, J.-M. Lasry, P.-L. Lions, H. Pham, E. Taflin, Paris-Princeton Lectures on Mathematical Finance 2004. Editors: R.A. Carmona, E. Çinlar, I. Ekeland, E. Jouini, J.A. Scheinkman, N. Touzi (2007)

Vol. 1920: S.N. Evans, Probability and Real Trees. Ecole d'Été de Probabilités de Saint-Flour XXXV-2005 (2008)

Vol. 1921: J.P. Tian, Evolution Algebras and their Applications (2008)

Vol. 1922: A. Friedman (Ed.), Tutorials in Mathematical BioSciences IV. Evolution and Ecology (2008)

Vol. 1923: J.P.N. Bishwal, Parameter Estimation in Stochastic Differential Equations (2008)

Vol. 1924: M. Wilson, Littlewood-Paley Theory and Exponential-Square Integrability (2008)

Vol. 1925: M. du Sautoy, L. Woodward, Zeta Functions of Groups and Rings (2008)

Vol. 1926: L. Barreira, V. Claudia, Stability of Nonautonomous Differential Equations (2008)

Vol. 1927: L. Ambrosio, L. Caffarelli, M.G. Crandall, L.C. Evans, N. Fusco, Calculus of Variations and Non-Linear Partial Differential Equations. Cetraro, Italy 2005. Editors: B. Dacorogna, P. Marcellini (2008)

Vol. 1928: J. Jonsson, Simplicial Complexes of Graphs (2008)

Vol. 1929: Y. Mishura, Stochastic Calculus for Fractional Brownian Motion and Related Processes (2008)

Vol. 1930: J.M. Urbano, The Method of Intrinsic Scaling. A Systematic Approach to Regularity for Degenerate and Singular PDEs (2008)

Vol. 1931: M. Cowling, E. Frenkel, M. Kashiwara, A. Valette, D.A. Vogan, Jr., N.R. Wallach, Representation Theory and Complex Analysis. Venice, Italy 2004. Editors: E.C. Tarabusi, A. D'Agnolo, M. Picardello (2008)

Vol. 1932: A.A. Agrachev, A.S. Morse, E.D. Sontag, H.J. Sussmann, V.I. Utkin, Nonlinear and Optimal Control Theory. Cetraro, Italy 2004. Editors: P. Nistri, G. Stefani (2008)

Vol. 1933: M. Petković, Point Estimation of Root Finding Methods (2008)

Vol. 1934: C. Donati-Martin, M. Émery, A. Rouault, C. Stricker (Eds.), Séminaire de Probabilités XLI (2008)

Vol. 1935: A. Unterberger, Alternative Pseudodifferential Analysis (2008)

Vol. 1936: P. Magal, S. Ruan (Eds.), Structured Population Models in Biology and Epidemiology (2008)

Vol. 1937: G. Capriz, P. Giovine, P.M. Mariano (Eds.), Mathematical Models of Granular Matter (2008)

Vol. 1938: D. Auroux, F. Catanese, M. Manetti, P. Seidel, B. Siebert, I. Smith, G. Tian, Symplectic 4-Manifolds and Algebraic Surfaces. Cetraro, Italy 2003. Editors: F. Catanese, G. Tian (2008)

Vol. 1939: D. Boffi, F. Brezzi, L. Demkowicz, R.G. Durán, R.S. Falk, M. Fortin, Mixed Finite Elements, Compatibility Conditions, and Applications. Cetraro, Italy 2006. Editors: D. Boffi, L. Gastaldi (2008)

Vol. 1940: J. Banasiak, V. Capasso, M.A.J. Chaplain, M. Lachowicz, J. Miękisz, Multiscale Problems in the Life Sciences. From Microscopic to Macroscopic. Będlewo, Poland 2006. Editors: V. Capasso, M. Lachowicz (2008)

Vol. 1941: S.M.J. Haran, Arithmetical Investigations. Representation Theory, Orthogonal Polynomials, and Quantum Interpolations (2008)

Vol. 1942: S. Albeverio, F. Flandoli, Y.G. Sinai, SPDE in Hydrodynamic. Recent Progress and Prospects. Cetraro, Italy 2005. Editors: G. Da Prato, M. Röckner (2008)

Vol. 1943: L.L. Bonilla (Ed.), Inverse Problems and Imaging. Martina Franca, Italy 2002 (2008)

Vol. 1944: A. Di Bartolo, G. Falcone, P. Plaumann, K. Strambach, Algebraic Groups and Lie Groups with Few Factors (2008)

Vol. 1945: F. Brauer, P. van den Driessche, J. Wu (Eds.), Mathematical Epidemiology (2008)

Vol. 1946: G. Allaire, A. Arnold, P. Degond, T.Y. Hou, Quantum Transport. Modelling Analysis and Asymptotics. Cetraro, Italy 2006. Editors: N.B. Abdallah, G. Frosali (2008)

Vol. 1947: D. Abramovich, M. Marino, M. Thaddeus, R. Vakil, Enumerative Invariants in Algebraic Geometry and String Theory. Cetraro, Italy 2005. Editors: K. Behrend, M. Manetti (2008)

Vol. 1948: F. Cao, J-L. Lisani, J-M. Morel, P. Musé, F. Sur, A Theory of Shape Identification (2008)

Vol. 1949: H.G. Feichtinger, B. Helffer, M.P. Lamoureux, N. Lerner, J. Toft, Pseudo-differential Operators. Cetraro, Italy 2006. Editors: L. Rodino, M.W. Wong (2008)

Vol. 1950: M. Bramson, Stability of Queueing Networks, Ecole d' Eté de Probabilités de Saint-Flour XXXVI-2006 (2008)

Recent Reprints and New Editions

Vol. 1702: J. Ma, J. Yong, Forward-Backward Stochastic Differential Equations and their Applications. 1999 – Corr. 3rd printing (2007)

Vol. 830: J.A. Green, Polynomial Representations of GL_n, with an Appendix on Schensted Correspondence and Littelmann Paths by K. Erdmann, J.A. Green and M. Schoker 1980 – 2nd corr. and augmented edition (2007)

Vol. 1693: S. Simons, From Hahn-Banach to Monotonicity (Minimax and Monotonicity 1998) – 2nd exp. edition (2008)

Vol. 470: R.E. Bowen, Equilibrium States and the Ergodic Theory of Anosov Diffeomorphisms. With a preface by D. Ruelle. Edited by J.-R. Chazottes. 1975 – 2nd rev. edition (2008)

Vol. 523: S.A. Albeverio, R.J. Høegh-Krohn, S. Mazzucchi, Mathematical Theory of Feynman Path Integral. 1976 – 2nd corr. and enlarged edition (2008)

Vol. 1764: A. Cannas da Silva, Lectures on Symplectic Geometry 2001 – Corr. 2nd printing (2008)